Hot topics in Cardio-Oncology

RIVER PUBLISHERS SERIES IN BIOTECHNOLOGY AND MEDICAL TECHNOLOGY FORUM

Series Editors

Indexing: all books published in this series are submitted to the Web of Science Book Citation Index (BkCI) and to SCOPUS for evaluation and indexing

Biotechnology & Medical Technology (BTMT) Forum is an international initiative aimed at disseminating the results of studies interfacing basic and translational medicine, biomaterials and applied engineering. BTMT Forum is committed to rapidly communicate the results of the scientific studies to scientists and the wider public worldwide.

Peer-reviewed original comprehensive articles are published monthly on the basis of their novelty, relevance, interdisciplinary impact, and potential scientific and translational significance. Early outcomes from clinical trials are also published. The Forum publishes thorough authoritative reviews and commentaries on the most important issues and perspectives of cutting-edge research after a careful evaluation by the Editorial Board.

Monographic books are also printed on specific topics of relevant concern under invitation by the Editorial Board or after independent proposals by potential book editors.

Finally, BTMT Forum publishes a book series specifically intended for professionals eager to revitalize their acquaintance and students of different educational ages.

BTMT Forum activities also involve the promotion of Congresses, Seminars, etc.

Topics

- basic and translational medicine
- biomaterials
- applied engineering

For a list of other books in this series, visit www.riverpublishers.com

Hot topics in Cardio-Oncology

Editors

Valentina Mercurio
Department of Translational Medical Sciences,
Federico II University, Naples, Italy

Pasquale Pagliaro
Department of Clinical and Biological Sciences,
University of Turin, Italy

Claudia Penna
Department of Clinical and Biological Sciences,
University of Turin, Italy

Carlo Gabriele Tocchetti
Department of Translational Medical Sciences,
Interdepartmental Center of Clinical and Translational Research (CIRCET),
Interdepartmental Hypertension Research Center (CIRIAPA),
Federico II University, Naples, Italy

River Publishers

Published, sold and distributed by:
River Publishers
Alsbjergvej 10
9260 Gistrup
Denmark

www.riverpublishers.com

ISBN: 9788770226288 (Hardback)
9788770226271 (Ebook)

Contents

List of Figures

List of Tables

List of Contributors

Ameri Pietro, *Cardiovascular Disease Unit, IRCCS Italian Cardiovascular Network, IRCCS Ospedale Policlinico San Martino, Genoa, Italy; Department of Internal Medicine, University of Genova, Genoa, Italy*

Antonio Carannante, *Department of Translational Medical Sciences, Federico II University, Naples, Italy*

Bruno Francesco, *Division of Cardiology, Department of Medical Science, Città della Salute e della Scienza, University of Turin, Italy; Division of Cardiac Surgery, Città della Salute e della Scienza, University of Turin, Italy*

Canonico Mario Enrico, *Department of Advanced Biomedical Science, Federico II University Hospital, Naples, Italy*

Castagno Davide, *Division of Cardiology, Department of Medical Sciences, "Città della Salute e della Scienza Hospital," University of Turin, Italy*

Cuomo Alessandra, *Department of Translational Medical Sciences, Federico II University, Naples, Italy*

Cusenza Vincenzo, *Division of Cardiology, Department of Medical Sciences, "Città della Salute e della Scienza Hospital," University of Turin, Italy*

De Ferrari Gaetano Maria, *Division of Cardiology, Department of Medical Sciences, "Città della Salute e della Scienza Hospital," University of Turin, Italy*

De Filippo Ovidio, *Division of Cardiology, Department of Medical Science, Città della Salute e della Scienza, University of Turin, Italy; Division of Cardiac Surgery, Città della Salute e della Scienza, University of Turin, Italy*

Deidda Martino, *Department of Medical Sciences and Public Health – University of Cagliari, Italy*

Dessalvi Christian Cadeddu, *Department of Medical Sciences and Public Health – University of Cagliari, Italy*

D'Angelo Giovanni, *Department of Translational Medical Sciences, Federico II University, Naples, Italy*

D'Ascenzo Fabrizio, *Division of Cardiology, Department of Medical Science, Città della Salute e della Scienza, University of Turin, Italy; Division of Cardiac Surgery, Città della Salute e della Scienza, University of Turin, Italy*

Galdiero Maria Rosaria, *Department of Translational Medical Sciences, Federico II University, Naples, Italy; Center for Basic and Clinical Immunology Research (CISI), Federico II University, Naples, Italy; WAO Center of Excellence, Naples, Italy; Institute of Experimental Endocrinology and Oncology "G. Salvatore" (IEOS), National Research Council (CNR), Naples, Italy*

Gallone Guglielmo, *Division of Cardiology, Department of Medical Science, Città della Salute e della Scienza, University of Turin, Italy; Division of Cardiac Surgery, Città della Salute e della Scienza, University of Turin, Italy*

Lisi Daniela Di, *Department of Health Promotion, Mother and Child Care, Internal Medicine and Medical Specialties, University of Palermo, Cardiology Unit, University Hospital P. Giaccone, Palermo, Italy*

Manganaro Roberta, *Department of Clinical and Experimental Medicine, Unit of Cardiology – University of Messina, Messina Italy*

Marone Giancarlo, *Department of Public Health, Section of Hygiene, University of Naples Federico II, Naples, Italy; Monaldi Hospital Pharmacy, Naples, Italy*

Mercurio Valentina, *Department of Translational Medical Sciences, Federico II University, Naples, Italy*

Mercuro Giuseppe, *Department of Medical Sciences and Public Health – University of Cagliari, Italy*

Noto Antonio, *Department of Medical Sciences and Public Health – University of Cagliari, Italy*

Novo Giuseppina, *Department of Health Promotion, Mother and Child Care, Internal Medicine and Medical Specialties, University of Palermo, Cardiology Unit, University Hospital P. Giaccone, Palermo, Italy*

Pagliaro Pasquale, *Department of Clinical and Biological Sciences, University of Turin, Torino, Italy; Istituto Nazionale per le Ricerche Cardiovascolari, Bologna, Italy*

Paudice Francesca, *Department of Translational Medical Sciences, Federico II University, Naples, Italy*

Penna Claudia, *Department of Clinical and Biological Sciences, University of Turin, Torino, Italy; Istituto Nazionale per le Ricerche Cardiovascolari, Bologna, Italy*

Perrotta Giovanni, *Department of Translational Medical Sciences, Federico II University, Naples, Italy*

Pirozzi Flora, *Department of Translational Medical Sciences, Federico II University, Naples, Italy*

Poto Remo, *Department of Translational Medical Sciences, Federico II University, Naples, Italy 1*

Rinaldi Mauro, *Division of Cardiology, Department of Medical Science, Città della Salute e della Scienza, University of Turin, Italy; Division of Cardiac Surgery, Città della Salute e della Scienza, University of Turin, Italy*

Salizzoni Stefano, *Division of Cardiology, Department of Medical Science, Città della Salute e della Scienza, University of Turin, Italy; Division of Cardiac Surgery, Città della Salute e della Scienza, University of Turin, Italy*

Santoro Ciro, *Department of Advanced Biomedical Science, Federico II University Hospital, Naples, Italy*

Spallarossa Paolo, *Clinic of Cardiovascular Diseases, IRCCS Ospedale Policlinico San Martino, Genova, Italy*

Tini Giacomo, *Cardiology, Azienda Ospedaliero-Universitaria Sant'Andrea, University of Rome Sapienza, Rome, Italy*

Tocchetti Carlo Gabriele, *Department of Translational Medical Sciences, Federico II University, Naples, Italy; Center for Basic and Clinical Immunology Research (CISI), Federico II University, Naples, Italy; Interdepartmental Center of Clinical and Translational Research (CIRCET), Federico II University, Naples, Italy; Interdepartmental Hypertension Research Center (CIRIAPA), Federico II University, Naples, Italy*

Varricchi Gilda, *Department of Translational Medical Sciences, Federico II University, Naples, Italy;Center for Basic and Clinical Immunology Research (CISI), Federico II University, Naples, Italy; WAO Center of Excellence, Naples, Italy; Institute of Experimental Endocrinology and Oncology "G. Salvatore" (IEOS), National Research Council (CNR), Naples, Italy*

Volpe Massimo, *Cardiology, Azienda Ospedaliero-Universitaria Sant'Andrea, University of Rome Sapienza, Rome, Italy*

Zito Concetta, *Department of Clinical and Experimental Medicine, Unit of Cardiology – University of Messina, Messina Italy*

List of Abbreviations

18F-FDG	18-fluorodeoxyglucose
1H NM	proton nuclear magnetic resonance spectroscopy
2-D Echo	two-dimensional echocardiography
2D	two-dimensional;
3D	three-dimensional;
3DE	three-dimensional echocardiography
5-FU	5-fluorouracil
AC	anthracyclines
ACEi	angiotensin converting enzyme-inihibitors
ACS	acute coronary syndrome
AF	Atrial fibrillation
Akt	
AMI	acute myocardial infarction
AMPK	Adenosine monophosphate-activated protein kinase
ANP	atrial natriuretic peptide
AOEs	arterial occlusive events
AP-1	activator protein 1
APC	antigen-presenting cells
ARBs	angiotensin receptor blockers
AS	Aortic Stenosis
ATEs	arterial thromboembolic events
BAD	Bcl-2-associated death promoter
Bcr	breakpoint cluster region;
BNP	brain natriuretic peptide
BP	blood pressure
CAD	coronary artery disease
CANTOS trial	Canakinumab Anti-Inflammatory Thrombosis Outcome Study trial
CAR-T	cell therapy uses T cells
CCB	calcium channel blockers
CHARM	Candesartan in Heart Failure: Assessment of Mortality and Morbidity trial

CK	creatine kinase
CK-MB	myocardial CK
CMP	cardiomyopathy;
CMR	cardiac magnetic resonance
CREB	cAMP response element-binding protein
CRP	C-reactive protein
CRS	cytokine release syndrome
CSF-1R	colony-stimulating factor-1 receptor
CT	computed tomography
CTLA- 4	Cytotoxic-T-lymphocyte-associated antigen 4
cTn	Cardiac troponins
CTRCD	Cancer Therapy-Related Cardiac Dysfunction
CTX	cardiotoxicity
CV	cardiovascular;
CVD	Cardiovascular disease
CY	cyclophosphamide
DAMPs	danger-associated molecular patterns
DAPT	double antiplatelet therapy
DOXO	Doxorubicin
DSI	Doppler strain imaging
DZR	dexrazoxane
ECG	Electrocardiogram
EDV	end-diastolic volume;
EF	ejection fraction
EMT	epithelial to mesenchymal transition
eNOS	endothelial nitric oxide synthase
ESV	end-systolic volume;
FAC	fractional area change;
FDA	Food and Drug Administration
FLT3	FMS-like tyrosine kinase-3
F-THP	free THP
GC-MS	gas-chromatography coupled with mass spectrometry
GCS	global circumferential strain;
GISSI-HF trial	Gruppo Italiano per lo Studio della Sopravvivenza nella Insufficienza Cardiaca-Heart Failure
GLS	global longitudinal strain;
GnRH	gonadotropin-releasing hormone
H_2O_2	hydrogen peroxide
HER2	human epidermal growth factor receptor 2;
HF	heart failure

HFrEF	Heart Failure with reduced ejection fraction
hiPSC-CMs	human induced pluripotent stem cell-derived cardiomyocytes
HO•	hydroxyl radical
HR	hazard ratio
HR-MAS NMR	high-resolution magic-angle-spinning nuclear magnetic resonance techniques
HTN	hypertension
ICI	Immune Checkpoint Inhibitors
ICOS-1	International Cardio Oncology Society-One trial
IFN-γ	interferon-gamma
IKr	Potassium rapid inward rectifier
IL	interleukin
IL-6R	soluble receptor IL-6
iNOS	inducible nitric oxide synthase
irAEs	immune-related adverse events
IVC	inferior vena cava;
LC-MS	liquid chromatography mass spectrometry
LDH	lactate dehydrogenase
LGE	late gadolinium enhancement
LPC	lysophosphatidylcholine
L-THP	liposome powder THP
LV	left ventricular
LVEF	Left ventricle EF
LVEF	left ventricular ejection fraction;
mAbs	monoclonal antibodies
MAPKs	mitogen-activated protein kinases
MetS	metabolic syndrome
MI	myocardial infarction
MTX	methotrexate
NF- kB	nuclear factor-κB
NK	natural killer
NO	nitric oxide
pro-BNP	pro-B-Type natriuretic peptide
NT-proBNP	N-terminal portion-proBNP
O^{-2}	superoxide anion
PAD	peripheral arterial occlusive disease
PAD	peripheral artery disease;
PAH	pulmonary arterial hypertension
PCI	Percutaneous Coronary Interventions

PDGFR	Platelet-derived growth factor receptor
PGE2	prostaglandin E2
PI3Ks	Phosphoinositide 3-kinases
PKC	protein kinase C
PRADA	primary prevention trial with Candesartan trial
Akt/PKB	protein kinase B Akt/PKB
RAAS	renin-angiotensin-aldosterone system
RIHD	radiation-induced" heart disease
ROS	Reactive oxygen species
RT	radiotherapy
RV	right ventricular;
RVEF	right ventricular ejection fraction;
SD	male Sprague Dawley rat
SEER	Surveillance Epidemiology and End Results
SNS	sympathetic nervous system
SOLVD	Studies of Left Ventricular Dysfunction trial.
SpD	spinochrome D
SSFP	steady-state free precession sequence.
STE	speckle-tracking echocardiography
SUCCOUR	Strain Surveillance During Chemotherapy for Improving Cardiovascular Outcome trail
TAM	tumor-associated macrophage
TAPSE	tricuspid annular plane systolic excursion;
TAVI	transcatheter aortic valve implantations
TDI	Tissue Doppler Imaging
THP	anthracycline pirarubicin
TKI	tyrosine kinase inhibitors
TLR	toll-like receptor .
TNF	tumour necrosis factor
TR	tricuspid regurgitation.
TSP2	thrombospondin-2
UPLC–Q-TOF-MS	ultra-performance liquid chromatography–quadrupole time-of-flight mass spectrometry
VEGF	vascular endothelial growth factor
VEGFi	vascular endothelial growth factor inhibitor
VHD	valvular heart disease.
vWf	von Willebrand factor
βARs	β-adrenergic receptors

Introduction

Valentina Mercurio, MD, PhD, FISC[1]; Pasquale Pagliaro, MD, PhD[2,3], Claudia Penna, BSc, PhD[2,3]; Carlo G Tocchetti, MD, PhD, FHFA, FISC[1,4,5,6]

[1]Department of Translational Medical Sciences, Federico II University, Naples, Italy
[2]Departiment of Clinical and Biological Sciences, University of Turin, Torino, Italy
[3]Istituto Nazionale per le Ricerche Cardiovascolari, Bologna, Italy
[4]Center for Basic and Clinical Immunology Research (CISI), Federico II University, Naples, Italy
[5]Interdepartmental Center of Clinical and Translational Research (CIRCET), Federico II University, Naples, Italy
[6]Interdepartmental Hypertension Research Center (CIRIAPA), Federico II University, Naples, Italy

The growing advances in the field of cancer therapies are leading to a progressive decrease in mortality rates for several cancers. On the other side, such valuable therapies have shown a wide spectrum of cardiotoxicities. Indeed, the cardiovascular system represents a possible target of several antineoplastic drugs, with different manifestations: ischemic vasospasm, thromboembolism, hypertension, arrhythmias, and ventricular dysfunction, leading to heart failure. Asymptomatic reduction in left ventricular systolic function and heart failure are the most frequent complications of long-term oncological regimens. Such cardiotoxic manifestations can develop or persist even after recovery from cancer. Chemotherapy-induced cardiotoxicity becomes even more relevant when we consider that elderly patients may have more "opportunity" to develop cancer and may already present with pre-existing chronic diseases and cardiovascular comorbidities (such as hypertension, obesity, diabetes, dyslipidemia, but also coronary artery disease and heart failure) in which inflammation (better "inflammaging") and oxidative stress play an important role.

Several studies have investigated the possible mechanisms underlying the development of cardiovascular adverse events with anticancer therapies. For instance, it is well-known that anthracyclines can cause cardiotoxicity through an increase in oxidative damage in the myocardium. On the other hand, biological therapies led to the discovery of other possible mechanisms of cardiotoxicity: some inhibitors of biological pathways can also interfere with intracellular signaling with a key role in the cardiovascular system. Finally, in the last years, an increase in the incidence of myocarditis, especially with the use of immune checkpoint inhibitors, has been observed, suggesting the importance of immune-inflammatory mechanisms in the development of cardiotoxicity. Multiple intersections between thrombosis and cancer exist, and balanced antithrombotic therapies need to be considered. It is fundamental to detect cardiotoxicity as early as possible. In this regard, cardiovascular imaging, spanning from echocardiography to cardiac magnetic resonance, has acquired a fundamental role in the early detection and the monitoring of cardiotoxicity.

Finally, concerning the use of biomarkers for the detection of cardiotoxicity, the new -omics sciences are emerging in this context with promising results. In particular, metabolomics has provided the first encouraging results in animal models, and the first data in small clinical trials show the ability of metabolomics to identify patients who are undergoing significant early cardiovascular toxicity.

The chapters of this book emphasize the above-mentioned concepts, providing the readers with the latest advances and insights in the ever-expanding field of cardio-oncology.

1

Inflammation in Cardio-Oncology

Remo Poto, MD[1]; Giancarlo Marone, PharmaD[2,3]; Flora Pirozzi, MD, PhD[1]; Alessandra Cuomo, MD[1]; Antonio Carannante, MD[1]; Maria Rosaria Galdiero, MD, PhD[1,4,5,6]; Carlo G Tocchetti, MD, PhD, FHFA, FISC[1,4,7,8]; Valentina Mercurio, MD, PhD, FISC[1]; Gilda Varricchi, MD, PhD[1,4,5,6]

[1]Department of Translational Medical Sciences, Federico II University, Naples, Italy
[2]Department of Public Health, Section of Hygiene, University of Naples Federico II, Naples, Italy
[3]Monaldi Hospital Pharmacy, Naples, Italy
[4]Center for Basic and Clinical Immunology Research (CISI), Federico II University, Naples, Italy
[5]WAO Center of Excellence, Naples, Italy
[6]Institute of Experimental Endocrinology and Oncology "G. Salvatore" (IEOS), National Research Council (CNR), Naples, Italy
[7]Interdepartmental Center of Clinical and Translational Research (CIRCET), Federico II University, Naples, Italy
[8]Interdepartmental Hypertension Research Center (CIRIAPA), Federico II University, Naples, Italy

Correspondence to:
Carlo Gabriele Tocchetti, MD, PhD, FHFA, FISC,
UOS Gestione del paziente oncologico in Medicina Interna
Dipartimento di Scienze Mediche Traslazionali
Centro Interdipartimentale di Ricerca Clinica e Traslazionale (CIRCET)
Centro interdipertimentale di ricerca per l'Ipertensione Arteriosa e Patologie Associate (CIRIAPA)
Universita' degli Studi di Napoli Federico II, Via Pansini 5
80131 Napoli, ITALY,
Phone +39-081-746-2242, Fax +39-081-746-2246,
carlogabriele.tocchetti@unina.it

KEYWORDS: Cancer; Cardiovascular diseases; Cardio-oncology; Immune Checkpoint inhibitors; Inflammation.

1.1 Introduction

Inflammation contributes to the pathophysiological features of cardiovascular disease (CVD) and cancer and is involved in the initiation, progression, and poor prognosis of both diseases (Libby and Kobold 2019). With the prolongation of life expectancy, both CVD and cancer are rising in the elderly population, requiring a deeper understanding of their molecular mechanisms (de Boer et al. 2020; Cuomo et al. 2021). Indeed, aging is characterized by an increase in the prevalence of several chronic and degenerative diseases, such as cancer and CVDs, with the involvement of oxidative stress and cellular senescence (Mercurio et al. 2020). Immunosenescence has been associated with chronic low-grade inflammation referred to as *inflammaging*, which participates in the development of frailty, disability, cancer, and CV disease. The name inflammaging indicates a broad immune dysregulation in elderly people, in which persistently increased levels of circulating pro-inflammatory mediators [such as interleukin (IL)-1β, IL-6, tumor necrosis factor (TNF)-α] and of the biomarker C-reactive protein (CRP) are associated with a blunted immune response (Liberale et al. 2020).

The chronic low-grade pro-inflammatory state contributes to the activation of leukocytes, endothelial and vascular smooth muscle cells, thus accelerating vascular aging and atherosclerosis and leading to increased incidence of CVD. Therefore, inflammation has emerged as an independent CV risk factor and pathogenic contributor to CVD. Thanks to recent advances in pharmacological therapies, CV death and, in particular, sudden cardiac death (Shen et al. 2017; Conrad et al. 2019; Moliner et al. 2019) have been reduced among HF patients. However, this has led to an increased burden of comorbidities, including cancer (Tini et al. 2020). In parallel, the analogue improvement of oncological management and treatments has considerably decreased the mortality linked to several cancers, while concomitantly increasing the comorbidity burden of oncological elderly patients. In this context, cancer and CVDs share common systemic pathogenic inflammatory pathways and mechanisms (de Boer et al. 2020; Tocchetti et al. 2020).

1.2 Inflammation in Cardiac Injury and Repair

Importantly, inflammation plays a key role in cardiac injury and repair. During cardiac ischemia, several endogenous ligands that act as "danger signals", also called danger-associated molecular patterns (DAMPs), are released upon injury and modulate inflammation (Arslan et al. 2011). For instance, myocardial injury, through the exposure/release of cardiac antigens, triggers a persistent cardiac T-cell response with subsequent production of pro-inflammatory cytokines, such as TNF-α, IL-1, and IL-6, thus contributing to a self-perpetuating inflammatory state that underlies adverse tissue remodeling (Prabhu and Frangogiannis 2016; Frantz et al. 2018; de Boer et al. 2019). In particular, TNF-α is a potent activator of nuclear factor-κB (NF- κB), a primary mediator of inflammation in cancer (Sethi et al. 2008; Aggarwal et al. 2012). Downstream signaling includes the over-activation of mitogen-activated protein kinases (MAPKs) and the abnormal stimulation of NF-κB. Besides, the consequent overexpression of pro-inflammatory cytokine and chemokine genes triggers inflammatory cells and causes oxidative stress, thus leading to DNA damage and increase in the likelihood of malignant mutations and cancer incidence modifying the tissue microenvironment (White et al. 2013).

The inflammatory response also enables the induction of regenerative processes following acute myocardial injury, leading to heart failure (HF). Blockade of the pro-inflammatory cytokine IL-1β with canakinumab, an interleukin-1β neutralizing monoclonal antibody, was shown to significantly reduce the rate of recurrent CV events in patients with previous myocardial infarction (Canakinumab Anti-Inflammatory Thrombosis Outcome Study – CANTOS trial) (Ridker et al. 2017b).

The inflammatory response is also involved in cardiotoxicity from anticancer drugs (Tocchetti et al. 2019; Tocchetti et al. 2020, Figure 1.1). Doxorubicin-induced damage also leads to the upregulation of pro-inflammatory toll-like receptor 4 (TLR4) in macrophages (Wang et al. 2016), higher levels as TNF-α and IL-6 and reduced levels of the anti-inflammatory cytokine IL-10 (Pecoraro et al. 2016). Cardiac function was preserved, and survival improved in TLR2 knock-out mice after DOXO exposure compared to wild-types (Nozaki et al. 2004). DOXO also induces local modulators of inflammation and fibrosis, produced by both macrophages and fibroblasts.

In addition, increased production of the matricellular protein thrombospondin-2 (TSP2) is protective in mice treated with DOXO. Augmented myocyte damage in the absence of TSP-2 was linked with impaired activation of the Akt signaling pathway. Importantly, inhibition of Akt phosphorylation in cardiomyocytes reduced TSP-2 expression, revealing a single feedback loop between Akt and TSP-2 (van Almen et al. 2011). Furthermore, CCL2/CCR2-dependent recruitment of functional antigen-presenting cells (APC) into tumor tissue is an expected therapeutic effect of anthracyclines (Ma et al. 2014).

1.3 Immune Checkpoint Inhibitors

For decades, oncologists have been developing strategies to modulate inflammation in order to achieve therapeutic anticancer immune responses (Lesterhuis et al. 2011). Cancer immunotherapies with monoclonal antibodies (mAbs) against immune checkpoints (i.e., CTLA-4 and PD-1/ PD-L1) have revolutionized antineoplastic treatments and stand out as the biggest example of the involvement of inflammatory processes in cardio-oncology; see Figure 1.1.). Cytotoxic-T-lymphocyte-associated antigen 4 (CTLA-4), PD-1, and its ligand PD-L1 are crucial regulators of the immune response (Varricchi et al. 2017; Lyon et al. 2018; Hu et al. 2019).

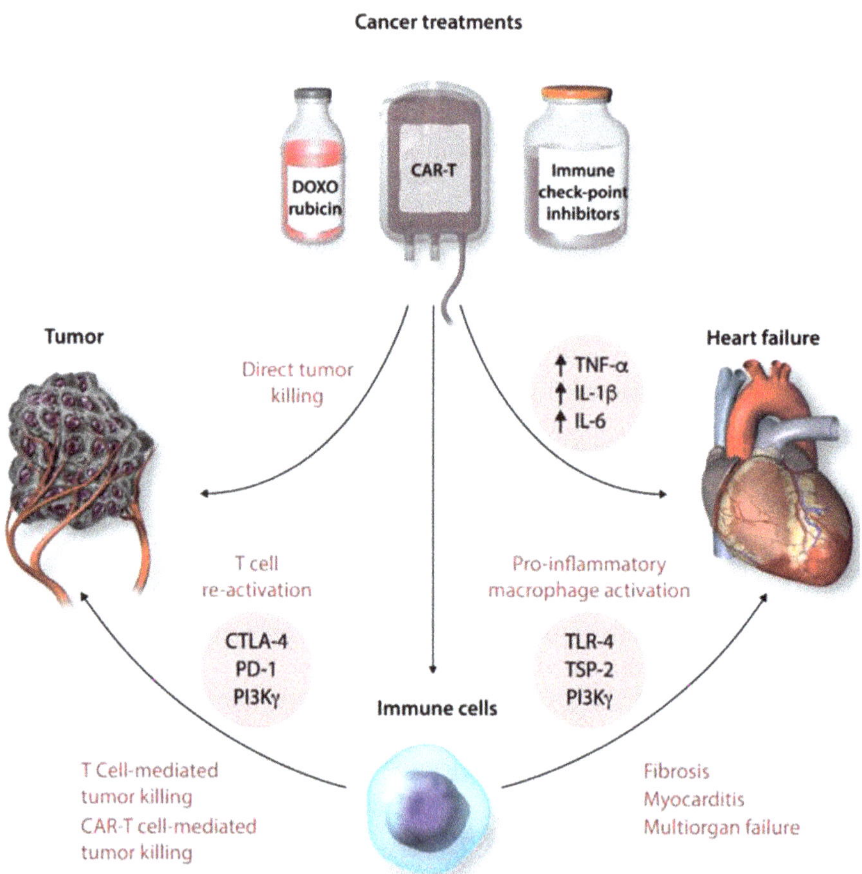

Figure 1.1 (Reproduced with permission from Tocchetti et al. 2020) Inflammation at the intersection of the anticancer action and cardiac side effects of major oncological treatments.

CTLA-4, a co-inhibitory molecule expressed on activated CD4+/CD8+ T-cells, competes with CD28 in binding CD80 and/or CD86 to attenuate T-cell activation (Linsley et al. 1990). Moreover, PD-1 expressed on T-cells, natural killer (NK) cells, B cells, monocytes, tumor-associated macrophage (TAM), immature Langerhans cells and cardiomyocytes, and its ligand PD-L1, inhibits the immune response by suppressing T-cell proliferation and reducing cytokine production (Swaika et al. 2015). The expression of PD-L1 is an essential immune evasion mechanism by which cancer cells escape the host immune response. Accordingly, several checkpoint inhibition strategies have been developed (Nguyen and Ohashi 2015; Sharma and Allison 2015, 2020; Le Mercier, Lines, and Noelle 2015). Monoclonal antibodies directed against CTLA-4 (ipilimumab), PD-1 (nivolumab, pembrolizumab, and cemiplimab), and PD-L1 (atezolizumab, avelumab, and durvalumab) block these immune checkpoints and unleash anti-tumor immunity, leading to tumor cell death through cytolytic molecules, such as TNF-α, granzyme B, and IFN-γ (Varricchi et al. 2017).

Immune checkpoints play a central role in the maintenance of self-tolerance (Tivol et al. 1995; Waterhouse et al. 1995; Nishimura et al. 1999; van Elsas et al. 2001; Fritz and Lenardo 2019). Inhibition of these pathways with ICIs, either alone or in combination, can lead to imbalances in immunologic tolerance that result in a broad spectrum of immune-related adverse events (irAEs) (Zimmer et al. 2016; Puzanov et al. 2017; Tocchetti et al. 2018). Cardiac irAEs due to ICIs (such as myocarditis, conduction abnormalities, cardiomyopathy, pericarditis, Takotsubo syndrome, and cardiac failure) are rare (Wang et al. 2017; Ederhy et al. 2018; Yang and Asnani 2018). The true incidence of cardiac irAEs due to ICIs is unknown; current estimates suggest less than 1% of patients (Johnson et al. 2016). Cardiac irAEs appear more frequently in patients treated with ipilimumab compared to PD-1 inhibitors (Johnson et al. 2016).

The pathophysiologic mechanisms of ICI-associated cardiotoxicity are still elusive and remain to be clarified. PD-1 and PD-L1 are expressed on the surface of murine and human cardiomyocytes (Dong et al. 1999; Freeman et al. 2000; Nishimura et al. 2001; Johnson et al. 2016). Experimental studies have demonstrated that CTLA-4 and PD-1 deletion or inhibition can cause autoimmune myocarditis with lymphocytic infiltration of cytotoxic T-cells (Tivol et al. 1995; Okazaki et al. 2003; Grabie et al. 2007; Wang et al. 2010;). Läubli *et al.* reported that lymphocytic infiltrates were characterized by the same T-cell lineage, which was present in both myocardium and tumor (Läubli et al. 2015). The latter finding suggests the existence of an expansion of a T-cell clone targeting a distinct but homologous antigen shared by the heart and the tumor. Johnson *et al.* (2016) found that T-cell clonal activation occurs in response to a common antigen, and the same T-cell clone identified within myocardium was also present in tumor and skeletal muscle.

Several factors may play a role in ICI-associated cardiotoxicity. In patients with CTLA-4 polymorphisms and concomitant cardiac risk factors, exposure to infectious agents can trigger a mechanism of molecular mimicry and can lead more easily to irAEs. Likely, in myocarditis patients, various stressors could play a contributing role (Giza et al. 2017). Gil-Cruz *et al.* (2019) demonstrated that progression of autoimmune myocarditis to lethal heart disease depends on cardiac myosin-specific Th17 cells imprinted in the intestine by a peptide mimic derived from a commensal *Bacteroides* species. Interestingly, they observed a significantly elevated *Bacteroides*-specific CD4+ T cell and B cell responses in human myocarditis. Furthermore, antibiotic therapy led to effective prevention of lethal disease in mice. The latter findings suggest that mimic peptides from commensal bacteria can promote inflammatory cardiomyopathy in genetically susceptible patients (Portig et al. 2009). It is possible to speculate that targeting the microbiome of genetically predisposed patients undergoing ICIs with antibiotics may reduce the incidence and severity of inflammatory cardiomyopathy (Gil-Cruz et al. 2019). This innovative approach is compatible with the increasing evidence that the inter-individual difference in gut microbiota is a source of the heterogeneity in immune-therapeutic efficacy and toxicity of ICIs (Vétizou et al. 2015; Zitvogel et al. 2018). Further studies in precision cardio-oncology are needed to elucidate individual and concurrent pathophysiologic mechanisms behind the cardiovascular toxicity of ICIs.

The holistic approach to cardiovascular management of patient candidates to ICI therapy requires a close collaboration among cardiologists, oncologists, and immunologists and is mandatory to stratify baseline risk and plan the best therapeutic approach for each patient (Lyon et al. 2018, 2020). Currently, there are no specific predictors of development of severe or mild cardiac toxicity caused by ICIs. Moreover, it is still unclear whether pre-existing risk factors might affect the incidence and severity of cardiac irAEs (Upadhrasta et al. 2019). Patients should be informed and alerted on potential development of ICI cardiotoxicity. These procedures represent a crucial component of preventive strategies before starting immunotherapy. Patients should be informed about signs and symptoms associated with cardiac toxicity and the need, especially in high-at-risk patients, for an in-depth screening and close surveillance (Hu et al. 2019; Lyon et al. 2018, 2020).

Prospective cardiovascular evaluation seems to be necessary to detect potential cardiotoxicity. Hence, screening with a higher level of vigilance is warranted before starting immunotherapy in patients with a known history of heart disease. It should be pointed out that novel strategies are needed to better identify high-at risk patients since conventional risk stratification algorithms such as Framingham risk score may underestimate cardiovascular risk in patients with cancer.

Pharmacological history should be assessed since many cancer patients have already received multiple cardiotoxic treatments before starting ICIs, such as anthracyclines (Mercurio et al. 2019), anti-ErbB2 drugs (de Lorenzo et al. 2018), RAF and MEK inhibitors (Heinzerling et al. 2019), tyrosine kinase inhibitors (Tocchetti et al. 2013; Sharma et al. 2017; Dobbin et al. 2018), and radiotherapy (Kirova et al. 2020). Importantly, these treatments can lead to the release/exposure of cardiac antigens with subsequent organ-specific immune responses, initially subclinical, which can be amplified by the administration of ICIs. Phenotyping both tumor antigen profiles and pre-treatment T-cell clones could contribute to baseline risk assessment (Lyon et al. 2018).

The assessment of pre-existing autoimmune diseases in cancer patients is an important issue. Previous clinical trials examining the safety and efficacy of ICIs have excluded patients with pre-existing autoimmune disorders. Therefore, the real incidence of flares of autoimmune disease cannot be estimated from these trials. Recent studies have started to evaluate the safety and efficacy of ICIs in patients with cancer and pre-existing autoimmune diseases. Tison *et al.* (2019) reported that 71% of patients experienced irAEs and 21% discontinued treatments. Inflammatory flares of the pre-existing autoimmune disease occurred in 47% of patients. Salem *et al.* (2018) reported that males appear to be at higher risk of developing cardiovascular events compared to females. This apparent sex dysmorphism could be explained by the fact that women have a higher incidence of autoimmune diseases and lower risk of cardiovascular diseases. For this reason, they are often excluded from clinical trials with ICIs, underestimating the true incidence of cardiac toxicity in women. Sex differences on cardiac toxicity during ICI remain controversial; therefore, further studies on this association are needed (Varricchi et al. 2018).

Baseline measurement of circulating cardiovascular biomarkers (i.e., troponin I or T, brain natriuretic peptide [BNP] or N-terminal pro-B-Type natriuretic peptide [NT pro-BNP], total CK, CRP, and fasting lipid profile) is mandatory in patient candidate for ICIs (Pudil et al. 2020). Electrocardiogram (ECG) and two-dimensional echocardiography (2D Echo) are recommended in patients with history or symptoms of cardiovascular disease (Čelutkienė et al. 2020). Moreover, it is essential to optimize the cardiovascular therapy in all patients.

Precision cardio-oncology, through multi "-omics" techniques, will help identify high-at-risk patients and personalize the clinical approach (Brownet al. 2020). At present, the impact of these techniques on clinical practice remains unknown. Novel precise and cost-effective risk prediction tools are emerging, with the need to identify patients who will benefit from closer monitoring and improve therapy decisions of cardiac toxicities. In this context, identification of gene polymorphisms leading to deficiency or dysfunction of CTLA-4, PD-1, or PD-L1 could be suggested in each patient since they are associated with myocarditis or cardiomyopathy and other autoimmune conditions in human and mice (Wang et al. 2010; Chen et al. 2013; Song et al. 2013; Wang et al. 2017). Integration of multiple systems including genomics, transcriptomics, proteomics, metabolomics, microRNA, microbiomics, and environmentomics could identify a fingerprint/signature unique to each patient, profiling a specific individual risk of cardiotoxicity (Brown et al. 2020).

Radiotherapy and immunotherapy could have a synergic effect in developing adverse events. For instance, administration of ICIs in patients with lung cancer after thoracic radiotherapy might trigger pericardial diseases due to the release of potentially shared antigens recognized by T-cells. Salem *et al.* (2018) showed that pericarditis was over-reported in lung cancer patients treated with anti-PD-1 or anti-PD-L1 therapy and radiotherapy.

1.4 CAR-T Cell Therapy

More recently, engineered T cells with chimeric antigen receptors (CAR-T cells; Figure 1.1) have been approved by the U.S. Food and Drug Administration (FDA) as the first genetically modified autologous T-cell that targets CD-19. A CAR is a recombinant receptor protein that has been engineered to activate T cells upon recognition of a specific antigen, resulting in the killing of target cells. CARcell therapy uses T cells engineered with CARs for cancer therapy. CD19 CAR-T cell therapies represent a new breakthrough in the treatment of relapsing and refractory hematological malignancies (Park et al. 2016; Salter et al. 2018). CD19 is a suitable target for CAR-T cells because it is expressed by B-cell malignancies but not by normal essential tissues. However, this promising therapy is associated with serious, potentially life-threatening events (Park et al. 2016). In particular, cytokine release syndrome (CRS) and neurotoxicity are associated with CD19 CAR-T cell therapies (Brudno and Kochenderfer 2016; Neelapu et al. 2017; Schuster et al. 2017). CRS is a systemic inflammatory response due to the widespread release of inflammatory cytokines (IL-2, soluble IL-2Rα, interferon-gamma, IL-6, soluble IL-6R, and granulocyte-macrophage colony-stimulating factor) and chemokines by activated lymphocytes or myeloid cells. It is associated with the activation of CAR-T cells and can be characterized by fever, hypotension, capillary leak, coagulopathy, and severe organ dysfunction. Recent evidence suggested that CAR-T cell therapies are also associated with cardiovascular toxicity (Burstein et al. 2018; Fitzgerald et al. 2017). It is reasonable that cardiovascular toxicities observed during CD19 CAR-T cell therapy are a direct consequence of the CRS.

This novel approach is also currently being investigated to be applied to different fields of interest. Epstein and colleagues (Aghajanian et al. 2019) demonstrated that engineered CAR-T cells can be exploited to reduce cardiac fibrosis and restore function in a mouse model of hypertensive heart failure. The authors speculated that engineered T cells could be used to target non-cancer cells and investigated that cardiac fibroblasts, which contribute to fibrosis following heart injury, could be effectively targeted by CAR-T cells. These results provide proof of concept for the possibility of treating cardiac fibrosis with engineered T cells.

1.5 Inflammation at the Crossroad Between Cancer and Cardiovascular Diseases

In cancer, inflammation plays a dual role. On the one hand, it has anti-tumorigenic function through the recognition and destruction of cancer cells; on the other hand, it predisposes to the development of cancer and promotes all stages of tumorigenesis, from initiation and promotion to invasion and metastasis (Galdiero et al. 2013). Recent evidence emphasized the involvement of several inflammatory mediators in the EMT (epithelial to mesenchymal transition), which is a crucial step toward tumor progression and malignant transformation, endowing the incipient cancer cell with invasive and metastatic properties (Chaffer and Weinberg 2011; Libby and Kobold 2019). Tumor-associated inflammation can be induced at different time points of tumor development. It can precede carcinogenesis in the form of autoimmunity or infection, can be hampered by malignant cells, or can be induced by anticancer therapy (Mantovani et al. 2008). In this way, novel strategies have emerged in the last decades to modulate inflammation in order to achieve therapeutic anticancer immune responses.

Visseren and colleagues of the UCC-SMART study group suggest that there is evidence that low-grade inflammation is also related to a higher risk of cancer (Van't Klooster et al. 2019). The authors reported that chronic systemic low-grade inflammation, measured by CRP levels <_10 mg/L, is a risk factor for incident cancer, markedly lung cancer, in patients with stable CVD. However, the relationship between inflammation and cancer is seen in former and current smokers and is uncertain in never smokers. By blocking the pro-inflammatory cytokine IL-1β, canakinumab reduced the rate of recurrent CV events significantly in patients with previous MI (CANTOS trial). Interestingly, it also seems that IL-1β blockade can protect from lung cancer mortality (Ridker et al. 2017a, 2017b). When administered to mice, DOXO induced an increase in serum IL-1β and other inflammatory factors (Sauter et al. 2011). Moreover, the IL-1β receptor antagonism protects against DOXO cardiotoxicity (Zhu et al. 2010). Similarly, inhibiting IL-6 with tocilizumab can protect against major adverse cardiovascular events MACE in patients treated with CAR-T (Alvi et al. 2019).

The experience of IL-1β blockade emphasizes that the identification of key molecules of the inflammatory pathways is important to halt both cancer and heart disease. α-1-antichymotrypsin, a member of the serine protease inhibitor (serpin) family of acute-phase proteins, is strongly associated with chronic systemic inflammation and is elevated in patients with heart failure (Bode et al. 2012; Surinova et al. 2015; Meijers et al. 2018). Meijers and colleagues also demonstraded that α-1-antichymotrypsin promotes colorectal cancer cell (HT-29) proliferation and tumor growth via protein kinase B pathways. The authors reported critical phosphorylation of protein kinase B and ribosomal protein S6 in HT-29 treated with α-1-antichymotrypsin (Meijers et al. 2018).

In addition to cytokines and chemokines, several lipid mediators such as prostanoids are involved in inflammatory pathways, but their role in cancer and CV disease is still elusive. Interestingly, prostaglandin E2 (PGE2) levels are elevated in cancer, particularly in gastrointestinal malignancies, and this prostaglandin promotes tumorigenesis and suppresses the immune response directed against cancer cells (Wang and DuBois 2018). Prostaglandin E2 can also affect cardiac function by decreasing the contractility of cardiomyocytes via prostaglandin EP receptors EP3, while, on the other hand, cardiomyocytes secrete cytokines with ability to induce chemotaxis (Gu et al. 2016). Prostanoids play a prominent role in the treatment of pulmonary arterial hypertension (PAH). A pre-clinical study showed that PGE2 promotes lung cancer migration (Kim et al. 2010). Another study revealed that prostacyclin prevents lung cancer in a mouse model (Nemenoff et al. 2008). However, cancer incidence in patients with PAH treated with prostaglandins or analogues is still unknown.

Along with these molecules, PI3Ks play an interesting role. In particular, the PI3Kγ isoform is augmented in both leukocytes and cardiomyocytes and not only control cardiomyocyte pathobiology but also modulate inflammatory pathways associated to different types of CV injury (Ghigo et al. 2010). PI3Kγ is upregulated in patients and mouse models of atherosclerosis and promotes leucocyte infiltration of the arterial wall, which is a crucial pathogenic mechanism in atherosclerosis (Fougerat et al. 2008).

PI3Kγ-mediated inflammation is also important for the cardiac response to pressure overload (Damilano et al. 2011). Moreover, by modulating the cardiac response to stress, macrophage PI3Kγ expression critically contributes to tumor growth and progression. Interestingly, macrophages participate with opposite roles in cancer and non-oncological inflammatory conditions. In response to pathogens or stimuli, macrophages express pro-inflammatory cytokines that stimulate cytotoxic T lymphocytes that eliminate pathogens and promote tissue repair. Contrariwise, in cancer macrophages express anti-inflammatory cytokines that lead to immune suppression, inhibit T cell-mediated tumor killing and ultimately promote resistance to immunotherapies (e.g., T-cell checkpoint inhibitors). PI3Kγ has been recently proposed as the molecular switch controlling immune stimulation and suppression during inflammation and cancer (Kaneda et al. 2016). With its dual role in both cancer and heart disease, PI3Kγ can be targeted pharmacologically to fight cancer and, at the same time, treat the heart (Li et al. 2018). This is particularly important for cancer patients treated with chemotherapy and suffering from iatrogenic cardiotoxicity (Kaneda et al. 2016; Skelly et al. 2018; Gangadhara et al. 2019). Further studies will assess this intriguing combined anticancer effect of such molecules in the context of cardiac protection.

Besides directly killing tumor cells, doxorubicin triggers cardiac inflammation via activation of macrophages and fibroblasts and the ensuing release of local modulators of inflammation and fibrosis, such as TNF-a, IL-1b, and IL-6. Major players of the inflammatory response induced by doxorubicin include macrophage TLR-4, the matricellular protein thrombospondin-2 (TSP-2), and leucocyte PI3Kγ. On the other hand, immune check point inhibitors (ICIs) inhibit molecules, such as cytotoxic-T-lymphocyte-associated antigen 4 (CTLA-4), programmed cell death 1 (PD-1), and its ligand PD-L1. As a consequence, anti-tumor immune cell responses are reactivated and lead to tumor cell death but concomitantly drive myocarditis. Although these new immunotherapies have notable anticancer effects, multiple mechanisms of immune resistance exist, and these might be overcome by using PI3Kγ inhibitors that reshape the tumor immune microenvironment. Finally, engineered T cells with chimeric antigen receptors (CAR-T cells) boost T-cell-mediated tumor killing but are burdened by cytokine release syndrome (CRS) leading to extremely serious complications, including cardiac and vascular dysfunction, and ultimately to multi-organ failure.

References

1. Aggarwal BB, Gupta SC, Kim JH. Historical perspectives on tumor necrosis factor and its superfamily: 25 years later, a golden journey. Blood. 2012;119:651–65.
2. Aghajanian H, Kimura T, Rurik JG, et al. Targeting cardiac fibrosis with engineered T cells. Nature. 2019;573:430–433. Correction Nature. 2019;576:E2.
3. van Almen GC, Swinnen M, Carai P, et al. Absence of thrombospondin-2 increases cardiomyocyte damage and matrix disruption in doxorubicin-induced cardiomyopathy. J Mol Cell Cardiol. 2011;51:318–28.
4. Alvi RM, Frigault MJ, Fradley MG, et al. Cardiovascular Events Among Adults Treated With Chimeric Antigen Receptor T-Cells (CAR-T). J Am Coll Cardiol. 2019;74:3099–3108.
5. Arslan F, de Kleijn DP, Pasterkamp G. Innate immune signaling in cardiac ischemia. Nat Rev Cardiol. 2011;8:292–300.
6. Bode JG, Albrecht U, Häussinger D, et al. Hepatic acute phase proteins--regulation by IL-6- and IL-1-type cytokines involving STAT3 and its crosstalk with NF-κB-dependent signaling. Eur J Cell Biol. 2012;91:496–505.
7. de Boer RA, Hulot JS, Tocchetti CG, et al. Common mechanistic pathways in cancer and heart failure. A scientific roadmap on behalf of the Translational Research Committee of the Heart Failure Association (HFA) of the European Society of Cardiology (ESC). Eur J Heart Fail. 2020;22:2272–2289.
8. de Boer RA, De Keulenaer G, Bauersachs J, et al. Towards better definition, quantification and treatment of fibrosis in heart failure. A scientific roadmap by the Committee of Translational Research of the Heart Failure Association (HFA) of the European Society of Cardiology. Eur J Heart Fail. 2019;21:272–285.
9. Brown SA, Ray JC, Herrmann J. Precision Cardio-Oncology: a Systems-Based Perspective on Cardiotoxicity of Tyrosine Kinase Inhibitors and Immune Checkpoint Inhibitors. J Cardiovasc Transl Res. 2020;13:402–416.
10. Brudno JN, Kochenderfer JN. Toxicities of chimeric antigen receptor T cells: recognition and management. Blood. 2016;127:3321–30.

11. Burstein DS, Maude S, Grupp S, et al. Cardiac Profile of Chimeric Antigen Receptor T Cell Therapy in Children: A Single-Institution Experience. Biol Blood Marrow Transplant. 2018;24:1590–1595.
12. Čelutkienė J, Lainscak M, Anderson L, et al. Imaging in patients with suspected acute heart failure: timeline approach position statement on behalf of the Heart Failure Association of the European Society of Cardiology. Eur J Heart Fail. 2020;22:181–195. Erratum in: Eur J Heart Fail. 2020;22:1287.
13. Chaffer CL, Weinberg RA. A perspective on cancer cell metastasis. Science. 2011;331:1559–64.
14. Chen Z, Fei M, Fu D, et al. Association between cytotoxic T lymphocyte antigen-4 polymorphism and type 1 diabetes: a meta-analysis. Gene. 2013;516:263–70.
15. Conrad N, Judge A, Canoy D, et al. Temporal Trends and Patterns in Mortality After Incident Heart Failure: A Longitudinal Analysis of 86 000 Individuals. JAMA Cardiol. 2019;4:1102–1111.
16. Cuomo A, Pirozzi F, Attanasio U, et al. Cancer Risk in the Heart Failure Population: Epidemiology, Mechanisms, and Clinical Implications. Curr Oncol Rep. 2020;23:7.
17. Damilano F, Franco I, Perrino C, et al. Distinct effects of leukocyte and cardiac phosphoinositide 3-kinase γ activity in pressure overload-induced cardiac failure. Circulation. 2011;123:391–9.
18. Dobbin SJH, Cameron AC, Petrie MC, et al. Toxicity of cancer therapy: what the cardiologist needs to know about angiogenesis inhibitors. Heart. 2018;104:1995–2002.
19. Dong H, Zhu G, Tamada K, et al. B7-H1, a third member of the B7 family, co-stimulates T-cell proliferation and interleukin-10 secretion. Nat Med. 1999;5:1365–9.
20. Ederhy S, Cautela J, Ancedy Y, et al. Takotsubo-Like Syndrome in Cancer Patients Treated With Immune Checkpoint Inhibitors. JACC Cardiovasc Imaging. 2018;11:1187–1190.
21. van Elsas A, Sutmuller RP, Hurwitz AA, et al. Elucidating the autoimmune and antitumor effector mechanisms of a treatment based on cytotoxic T lymphocyte antigen-4 blockade in combination with a B16 melanoma vaccine: comparison of prophylaxis and therapy. J Exp Med. 2001;194:481–9.
22. Fitzgerald JC, Weiss SL, Maude SL, et al. Cytokine Release Syndrome After Chimeric Antigen Receptor T Cell Therapy for Acute Lymphoblastic Leukemia. Crit Care Med. 2017;45:e124-e131.

23. Fougerat A, Gayral S, Gourdy P, et al. Genetic and pharmacological targeting of phosphoinositide 3-kinase-gamma reduces atherosclerosis and favors plaque stability by modulating inflammatory processes. Circulation. 2008;117:1310–7.
24. Frantz S, Falcao-Pires I, Balligand JL, et al. The innate immune system in chronic cardiomyopathy: a European Society of Cardiology (ESC) scientific statement from the Working Group on Myocardial Function of the ESC. Eur J Heart Fail. 2018;20:445–459.
25. Freeman GJ, Long AJ, Iwai Y, et al. Engagement of the PD-1 immunoinhibitory receptor by a novel B7 family member leads to negative regulation of lymphocyte activation. J Exp Med. 2000;192:1027–34.
26. Fritz JM, Lenardo MJ. Development of immune checkpoint therapy for cancer. J Exp Med. 2019;216:1244–1254.
27. Galdiero MR, Garlanda C, Jaillon S, et al. Tumor associated macrophages and neutrophils in tumor progression. J Cell Physiol. 2013;228: 1404–12.
28. Gangadhara G, Dahl G, Bohnacker T, et al. A class of highly selective inhibitors bind to an active state of PI3Kγ. Nat Chem Biol. 2019;15:348–357.
29. Ghigo A, Damilano F, Braccini L, et al. PI3K inhibition in inflammation: Toward tailored therapies for specific diseases. Bioessays. 2010;32:185–96.
30. Gil-Cruz C, Perez-Shibayama C, De Martin A, et al. Microbiota-derived peptide mimics drive lethal inflammatory cardiomyopathy. Science. 2019;366:881–886.
31. Giza DE, Lopez-Mattei J, Vejpongsa P, et al. Stress-Induced Cardiomyopathy in Cancer Patients. Am J Cardiol. 2017;120:2284–2288
32. Grabie N, Gotsman I, DaCosta R, et al. Endothelial programmed death-1 ligand 1 (PD-L1) regulates CD8+ T-cell mediated injury in the heart. Circulation. 2007;116:2062–71.
33. Gu X, Xu J, Zhu L, et al. Prostaglandin E2 Reduces Cardiac Contractility via EP3 Receptor. Circ Heart Fail. 2016;9:10.1161/CIRCHEARTFAILURE.116.003291 e003291.
34. Heinzerling L, Eigentler TK, Fluck M, et al. Tolerability of BRAF/MEK inhibitor combinations: adverse event evaluation and management. ESMO Open. 2019;4:e000491.
35. Hu JR, Florido R, Lipson EJ, et al. Cardiovascular toxicities associated with immune checkpoint inhibitors. Cardiovasc Res. 2019;115:854–868. Erratum in: Cardiovasc Res. 2019;115:868.

36. Johnson DB, Balko JM, Compton ML, et al. Fulminant Myocarditis with Combination Immune Checkpoint Blockade. N Engl J Med. 2016;375:1749–1755.
37. Kaneda MM, Messer KS, Ralainirina N, et al. PI3Kγ is a molecular switch that controls immune suppression. Nature. 2016;539:437–442. Erratum in: Nature. 2017;542:124.
38. Kim JI, Lakshmikanthan V, Frilot N, et al Prostaglandin E2 promotes lung cancer cell migration via EP4-betaArrestin1-c-Src signalsome. Mol Cancer Res. 2010;8:569–77.
39. Kirova Y, Tallet A, Aznar MC, et al. Radio-induced cardiotoxicity: From physiopathology and risk factors to adaptation of radiotherapy treatment planning and recommended cardiac follow-up. Cancer Radiother. 2020;24:576–585.
40. LäubliH,BalmelliC,BossardM,etal.Acuteheartfailureduetoautoimmune myocarditis under pembrolizumab treatment for metastatic melanoma. J Immunother Cancer. 2015;3:11.
41. Lesterhuis WJ, Haanen JB, Punt CJ. Cancer immunotherapy--revisited. Nat Rev Drug Discov. 2011;10:591–600.
42. Li M, Sala V, De Santis MC, et al. Phosphoinositide 3-Kinase Gamma Inhibition Protects From Anthracycline Cardiotoxicity and Reduces Tumor Growth. Circulation. 2018;138:696–711.
43. Libby P, Kobold S. Inflammation: a common contributor to cancer, aging, and cardiovascular diseases-expanding the concept of cardio-oncology. Cardiovasc Res. 2019;115:824–829.
44. Liberale L, Montecucco F, Tardif JC, et al. Inflamm-ageing: the role of inflammation in age-dependent cardiovascular disease. Eur Heart J. 2020;41:2974–2982.
45. Linsley PS, Clark EA, Ledbetter JA. T-cell antigen CD28 mediates adhesion with B cells by interacting with activation antigen B7/BB-1. Proc Natl Acad Sci U S A. 1990;87:5031–5.
46. De Lorenzo C, Paciello R, Riccio G, et al. Cardiotoxic effects of the novel approved anti-ErbB2 agents and reverse cardioprotective effects of ranolazine. Onco Targets Ther. 2018;11:2241–2250
47. Lyon AR, Dent S, Stanway S, et al. Baseline cardiovascular risk assessment in cancer patients scheduled to receive cardiotoxic cancer therapies: a position statement and new risk assessment tools from the Cardio-Oncology Study Group of the Heart Failure Association of the European Society of Cardiology in collaboration with the International Cardio-Oncology Society. Eur J Heart Fail. 2020;22:1945–1960.

48. Lyon AR, Yousaf N, Battisti NML, et al. Immune checkpoint inhibitors and cardiovascular toxicity. Lancet Oncol. 2018;19:e447-e458.
49. Ma Y, Mattarollo SR, Adjemian S, et al. CCL2/CCR2-dependent recruitment of functional antigen-presenting cells into tumors upon chemotherapy. Cancer Res. 2014;74:436–45.
50. Mantovani A, Allavena P, Sica A, et al. Cancer-related inflammation. Nature. 2008;454:436–44.
51. Meijers WC, Maglione M, Bakker SJL, et al. Heart Failure Stimulates Tumor Growth by Circulating Factors. Circulation. 2018;138:678–691.
52. Le Mercier I, Lines JL, Noelle RJ. Beyond CTLA-4 and PD-1, the Generation Z of Negative Checkpoint Regulators. Front Immunol. 2015;6:418.
53. Mercurio V, Cuomo A, Cadeddu Dessalvi C, et al. Redox Imbalances in Ageing and Metabolic Alterations: Implications in Cancer and Cardiac Diseases. An Overview from the Working Group of Cardiotoxicity and Cardioprotection of the Italian Society of Cardiology (SIC). Antioxidants (Basel). 2020;9:641.
54. Mercurio V, Cuomo A, Della Pepa R, et al. What Is the Cardiac Impact of Chemotherapy and Subsequent Radiotherapy in Lymphoma Patients? Antioxid Redox Signal. 2019;31:1166–1174.
55. Moliner P, Lupón J, de Antonio M, et al. Trends in modes of death in heart failure over the last two decades: less sudden death but cancer deaths on the rise. Eur J Heart Fail. 2019;21:1259–1266.
56. Neelapu SS, Locke FL, Bartlett NL, et al. Axicabtagene Ciloleucel CAR T-Cell Therapy in Refractory Large B-Cell Lymphoma. N Engl J Med. 2017;377:2531–2544.
57. Nemenoff R, Meyer AM, Hudish TM, et al. Prostacyclin prevents murine lung cancer independent of the membrane receptor by activation of peroxisomal proliferator--activated receptor gamma. Cancer Prev Res (Phila). 2008;1:349–56.
58. Nguyen LT, Ohashi PS. Clinical blockade of PD1 and LAG3--potential mechanisms of action. Nat Rev Immunol. 2015;15:45–56.
59. Nishimura H, Okazaki T, Tanaka Y, et al. Autoimmune dilated cardiomyopathy in PD-1 receptor-deficient mice. Science. 2001;291:319–22.
60. Nishimura H, Nose M, Hiai H, et al. Development of lupus-like autoimmune diseases by disruption of the PD-1 gene encoding an ITIM motif-carrying immunoreceptor. Immunity. 1999;11:141–51.
61. Nozaki N, Shishido T, Takeishi Y, et al. Modulation of doxorubicin-induced cardiac dysfunction in toll-like receptor-2-knockout mice. Circulation. 2004;110:2869–74.

62. Okazaki T, Tanaka Y, Nishio R, et al. Autoantibodies against cardiac troponin I are responsible for dilated cardiomyopathy in PD-1-deficient mice. Nat Med. 2003;9:1477–83.
63. Park JH, Geyer MB, Brentjens RJ. CD19-targeted CAR T-cell therapeutics for hematologic malignancies: interpreting clinical outcomes to date. Blood. 2016;127:3312–20.
64. Pecoraro M, Del Pizzo M, Marzocco S, et al. Inflammatory mediators in a short-time mouse model of doxorubicin-induced cardiotoxicity. Toxicol Appl Pharmacol. 2016;293:44–52.
65. Portig I, Sandmoeller A, Kreilinger S, et al. HLA-DQB1* polymorphism and associations with dilated cardiomyopathy, inflammatory dilated cardiomyopathy and myocarditis. Autoimmunity. 2009;42:33–40.
66. Prabhu SD, Frangogiannis NG. The Biological Basis for Cardiac Repair After Myocardial Infarction: From Inflammation to Fibrosis. Circ Res. 2016;119:91–112.
67. Pudil R, Mueller C, Čelutkienė J, et al. Role of serum biomarkers in cancer patients receiving cardiotoxic cancer therapies: a position statement from the Cardio-Oncology Study Group of the Heart Failure Association and the Cardio-Oncology Council of the European Society of Cardiology. Eur J Heart Fail. 2020;22:1966–1983.
68. Puzanov I, Diab A, Abdallah K, et al., Society for Immunotherapy of Cancer Toxicity Management Working Group. Managing toxicities associated with immune checkpoint inhibitors: consensus recommendations from the Society for Immunotherapy of Cancer (SITC) Toxicity Management Working Group. J Immunother Cancer. 2017;5:95.
69. Ridker PM, Everett BM, Thuren T, et al; CANTOS Trial Group. Antiinflammatory Therapy with Canakinumab for Atherosclerotic Disease. N Engl J Med. 2017a;377:1119–1131.
70. Ridker PM, MacFadyen JG, Thuren T, et al.; CANTOS Trial Group. Effect of interleukin-1β inhibition with canakinumab on incident lung cancer in patients with atherosclerosis: exploratory results from a randomised, double-blind, placebo-controlled trial. Lancet. 2017b;390: 1833–1842.
71. Salem JE, Manouchehri A, Moey M, et al. Cardiovascular toxicities associated with immune checkpoint inhibitors: an observational, retrospective, pharmacovigilance study. Lancet Oncol. 2018;19:1579–1589.
72. Salter AI, Pont MJ, Riddell SR. Chimeric antigen receptor-modified T cells: CD19 and the road beyond. Blood. 2018;131:2621–2629.

73. Sauter KA, Wood LJ, Wong J, et al. Doxorubicin and daunorubicin induce processing and release of interleukin-1β through activation of the NLRP3 inflammasome. Cancer Biol Ther. 2011;11:1008–16.
74. Schuster SJ, Svoboda J, Chong EA, et al. Chimeric Antigen Receptor T Cells in Refractory B-Cell Lymphomas. N Engl J Med. 2017;377:2545–2554.
75. Sethi G, Sung B, Aggarwal BB. TNF: a master switch for inflammation to cancer. Front Biosci. 2008;13:5094–107.
76. Sharma A, Burridge PW, McKeithan WL, et al. High-throughput screening of tyrosine kinase inhibitor cardiotoxicity with human induced pluripotent stem cells. Sci Transl Med. 2017;9:eaaf2584.
77. Sharma P, Allison JP. The future of immune checkpoint therapy. Science. 2015;348:56–61.
78. Sharma P, Allison JP. Dissecting the mechanisms of immune checkpoint therapy. Nat Rev Immunol. 2020;20:75–76.
79. Shen L, Jhund PS, Petrie MC, et al. Declining Risk of Sudden Death in Heart Failure. N Engl J Med. 2017;377:41–51.
80. Skelly DA, Squiers GT, McLellan MA, et al. Single-Cell Transcriptional Profiling Reveals Cellular Diversity and Intercommunication in the Mouse Heart. Cell Rep. 2018;22:600–610.
81. Song GG, Kim JH, Kim YH, et al. Association between CTLA-4 polymorphisms and susceptibility to Celiac disease: a meta-analysis. Hum Immunol. 2013;74:1214–8.
82. Surinova S, Choi M, Tao S, et al. Prediction of colorectal cancer diagnosis based on circulating plasma proteins. EMBO Mol Med. 2015;7:1166–78.
83. Swaika A, Hammond WA, Joseph RW. Current state of anti-PD-L1 and anti-PD-1 agents in cancer therapy. Mol Immunol. 2015;67:4–17.
84. Tini G, Bertero E, Signori A, et al. Cancer Mortality in Trials of Heart Failure With Reduced Ejection Fraction: A Systematic Review and Meta-Analysis. J Am Heart Assoc. 2020;9:e016309.
85. Tison A, Quéré G, Misery L, et al; Groupe de Cancérologie Cutanée, Groupe Français de Pneumo-Cancérologie, and Club Rhumatismes et Inflammations. Safety and Efficacy of Immune Checkpoint Inhibitors in Patients With Cancer and Preexisting Autoimmune Disease: A Nationwide, Multicenter Cohort Study. Arthritis Rheumatol. 2019;71:2100–2111.
86. Tivol EA, Borriello F, Schweitzer AN, et al. Loss of CTLA-4 leads to massive lymphoproliferation and fatal multiorgan tissue destruction, revealing a critical negative regulatory role of CTLA-4. Immunity. 1995;3:541–7.

87. Tocchetti CG, Galdiero MR, Varricchi G. Cardiac Toxicity in Patients Treated With Immune Checkpoint Inhibitors: It Is Now Time for Cardio-Immuno-Oncology. J Am Coll Cardiol. 2018;71:1765–1767.
88. Tocchetti CG, Gallucci G, Coppola C, et al. The emerging issue of cardiac dysfunction induced by antineoplastic angiogenesis inhibitors. Eur J Heart Fail. 2013;15:482–9.
89. Tocchetti CG, Cadeddu C, Di Lisi D, Femminò S, Madonna R, Mele D, Monte I, Novo G, Penna C, Pepe A, Spallarossa P, Varricchi G, Zito C, Pagliaro P, Mercuro G.From Molecular Mechanisms to Clinical Management of Antineoplastic Drug-Induced Cardiovascular Toxicity: A Translational Overview. Antioxid Redox Signal. 2019 Jun 20;30(18):2110–2153. doi: 10.1089/ars.2016.6930. Epub 2017 May 15.PMID: 28398124
90. Tocchetti CG, Ameri P, de Boer RA, et al. Cardiac dysfunction in cancer patients: beyond direct cardiomyocyte damage of anticancer drugs: novel cardio-oncology insights from the joint 2019 meeting of the ESC Working Groups of Myocardial Function and Cellular Biology of the Heart. Cardiovasc Res. 2020;116:1820–1834.
91. Upadhrasta S, Elias H, Patel K, et al. Managing cardiotoxicity associated with immune checkpoint inhibitors. Chronic Dis Transl Med. 2019;5:6–14.
92. Van't Klooster CC, Ridker PM, Hjortnaes J, et al. The relation between systemic inflammation and incident cancer in patients with stable cardiovascular disease: a cohort study. Eur Heart J. 2019;40:3901–3909.
93. Varricchi G, Galdiero MR, Mercurio V, et al. Pharmacovigilating cardiotoxicity of immune checkpoint inhibitors. Lancet Oncol. 2018;19: 1545–1546.
94. Varricchi G, Galdiero MR, Tocchetti CG. Cardiac Toxicity of Immune Checkpoint Inhibitors: Cardio-Oncology Meets Immunology. Circulation. 2017;136:1989–1992.
95. Vétizou M, Pitt JM, Daillère R, et al. Anticancer immunotherapy by CTLA-4 blockade relies on the gut microbiota. Science. 2015;350:1079–84.
96. Wang DY, Okoye GD, Neilan TG, et al. Cardiovascular Toxicities Associated with Cancer Immunotherapies. Curr Cardiol Rep. 2017;19:21.
97. Wang D, DuBois RN. Role of prostanoids in gastrointestinal cancer. J Clin Invest. 2018;128:2732–2742.
98. Wang J, Okazaki IM, Yoshida T, et al. PD-1 deficiency results in the development of fatal myocarditis in MRL mice. Int Immunol. 2010;22:443–52.

99. Wang K, Zhu Q, Lu Y, et al. CTLA-4 +49 G/A Polymorphism Confers Autoimmune Disease Risk: An Updated Meta-Analysis. Genet Test Mol Biomarkers. 2017;21:222–227.
100. Wang L, Chen Q, Qi H, et al. Doxorubicin-Induced Systemic Inflammation Is Driven by Upregulation of Toll-Like Receptor TLR4 and Endotoxin Leakage. Cancer Res. 2016;76:6631–6642.
101. Waterhouse P, Penninger JM, Timms E, et al. Lymphoproliferative disorders with early lethality in mice deficient in Ctla-4. Science. 1995;270:985–8.
102. White GE, Iqbal AJ, Greaves DR. CC chemokine receptors and chronic inflammation--therapeutic opportunities and pharmacological challenges. Pharmacol Rev. 2013;65:47–89.
103. Yang S, Asnani A. Cardiotoxicities associated with immune checkpoint inhibitors. Curr Probl Cancer. 2018;42:422–432.
104. Zhu J, Zhang J, Xiang D, et al. Recombinant human interleukin-1 receptor antagonist protects mice against acute doxorubicin-induced cardiotoxicity. Eur J Pharmacol. 2010;643:247–53.
105. Zimmer L, Goldinger SM, Hofmann L, et al. Neurological, respiratory, musculoskeletal, cardiac and ocular side-effects of anti-PD-1 therapy. Eur J Cancer. 2016;60:210–25.
106. Zitvogel L, Ma Y, Raoult D, et al. The microbiome in cancer immunotherapy: Diagnostic tools and therapeutic strategies. Science. 2018;359:1366–1370.

2

Vascular Toxicity and Thromboembolic Risk in Cardio-Oncology

Daniela Di Lisi[1], Giuseppina Novo[1]

[1]Department of Health Promotion, Mother and Child Care, Internal Medicine and Medical Specialties, University of Palermo, Cardiology Unit, University Hospital P. Giaccone, Palermo, Italy.

****Correspondence to:***
Giuseppina Novo, MD, PhD., Chair of Cardiology, University of Palermo, Via Del Vespro 129, 90127 Palermo, Italy. Email: giuseppina.novo@unipa.it

KEYWORDS: Cancer; Cardiotoxicity; Cardiovascular Diseases; Hypertension; Risk Stratification.

2.1 Introduction

Chemotherapy-related vascular toxicity has a wide spectrum of clinical presentations: acute coronary syndrome (ACS), systemic hypertension, pulmonary arterial hypertension, stroke, and peripheral artery disease.

Especially agents that interfere with the vascular endothelial growth factor signaling pathway (VEGF – VEGFR inhibitors) can cause vascular side effects ranging from arterial hypertension to arterial events and cardiomyocyte toxicity.

Tyrosine kinase inhibitors targeting the Philadelphia chromosome mutation product (anti BCR-ABL) such as nilotinib and ponatinib have been associated with progressive atherosclerosis and acute vascular events.

Other drugs such as antimetabolites, anti-microtubule agents, and monoclonal antibodies can also cause vascular toxicity and increase the risk of vascular complications and the thromboembolic risk.

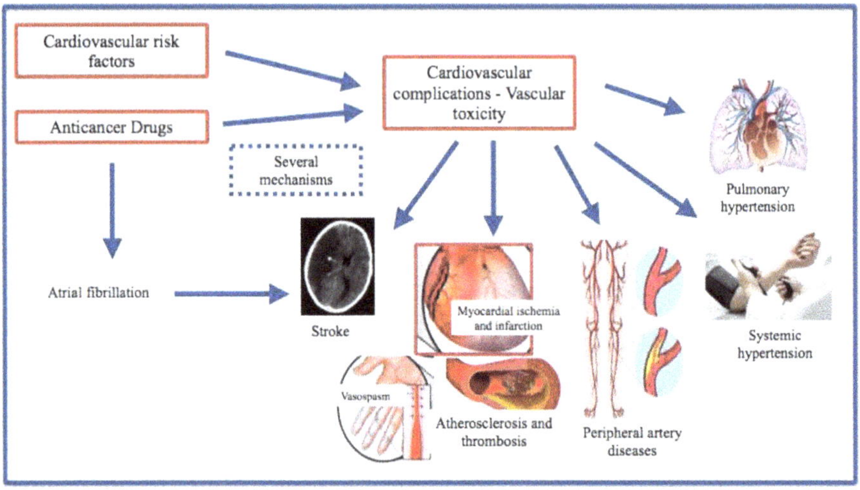

Figure 2.1 Vascular events caused by anticancer drugs.

Ibrutinib can cause atrial fibrillation, therefore increasing the thromboembolic risk especially in patients with other risk factors and high CHA2DS2 Vasc score.

Cancer itself with several mechanisms can contribute to increase the thromboembolic risk.

Thus, in Chapter II, we will describe vascular events caused by anticancer drugs, their mechanisms, and the factors that contribute to increase the thromboembolic risk in cancer patients (Figure 2.1).

2.2 Acute Coronary Syndrome (ACS) and Coronary Artery Diseases (CAD)

ACS can develop in cancer patients encompassing the entire spectrum from unstable angina to acute myocardial infarction (AMI). Several mechanisms have been hypothesized: vasospasm, acute endothelial damage, platelet–platelet activation, and aggregation and attraction of elevated low-density lipoprotein cholesterol particles (Giza et al. 2017). This can lead to formation of potentially unstable lipid-rich coronary plaques and to the initiation and acceleration of the atherosclerosis process (Davignon 2004). In addition, cancer cells can activate the coagulation cascade and other prothrombotic properties of host cells, inducing a prothrombotic state and provoking thrombosis (Razak et al. 2018).

5-fluorouracil (5-FU), capecitabine and gemcitabine can cause myocardial ischemia inducing vasospasm through several mechanisms (Cardinale et al. 2006; Lu et al. 2006; Ozturk 2009). For example, 5-FU can induce coronary spasm through a toxic effect on the vascular endothelium altering the endothelial nitric oxide synthase (eNOS) and causing endothelium-independent vasoconstriction via protein kinase C (Senkus et al. 2011). Nitric oxide (NO) produced by eNOS and its interaction with serine/threonine protein kinase Akt/PKB, caveolin, and calmodulin is a key determinant of cardiovascular tone (Chong and Ghosh 2019).

High plasma levels of endothelin-1 were observed in patients receiving 5-FU, especially in those experiencing cardiotoxicity. This observation supports the hypothesis of 5-FU-induced vasoconstriction (Thyss et al. 1992).

The incidence of myocardial ischemia with 5-FU and capecitabine is reported to be 3%–9%. Chest pain onset is often abrupt during infusion of 5-FU but can also be delayed, presenting within the first 72 hours after 5-FU administration (Van Cutsem et al. 2002).

Patients with previous documented coronary artery diseases (CAD) are especially vulnerable to high doses of 5-FU and are considered to have an increased risk of developing cardiac ischemic events. In addition to high doses of 5-FU, continuous infusion, prior mediastinal radiation, or simultaneous administration of other chemotherapeutic agents are factors that can contribute to the progress of cardiac ischemia in patients treated with antimetabolite drugs (Anand 1994). Conversely, the presence of cardiac risk factors does not appear to completely predict the development of adverse cardiac side effects (Kosmas et al. 2008).

Profound and prolonged vasospasm has also been implied in ACS presentations of **paclitaxel** (anti-microtubule agents), **rituximab** (monoclonal antibodies), and **sorafenib** (TKI) (Schrader et al. 2005; Armitage et al. 2008). The incidence of myocardial ischemia with sorafenib is around 2.7%–3% (Escudier et al. 2007; Arima et al. 2009).

Coronary spasm is due to hyper-contraction of coronary smooth muscle. Enhanced Ca^{2+} sensitization plays a critical role in the genesis of coronary spasm (Yasue et al. 2008). Some studies showed activation of small GTPase RhoA and its downstream effector, Rho-kinase (ROCK), leading to augmentation of Ca^{2+} sensitization. Sorafenib is a multikinase inhibitor that targets both Raf and vascular endothelial growth factor (VEGF) and platelet-derived growth factor receptor tyrosine kinase signaling. Its use downregulates MEK activity; therefore, it may cause upregulation of the Rho/ROCK pathway and augmentation of Ca^{2+} sensitization leading to the exacerbation of coronary artery spasm (Wilhelm et al. 2008). Similar mechanisms are involved in the pathophysiology of vasospasm caused by other drugs (Herrmann et al. 2016).

Diagnosis of coronary artery disease is performed by coronary angiography: when significant CAD and acute plaque rupture are ruled out, coronary artery vasospasm should be considered in patients treated with these drugs.

Other anticancer drugs can cause ACS and type I myocardial infarction as a consequence of the well-established types of plaque complications. Given the toxic effect of chemotherapeutics on the endothelial cells, there might be a greater propensity toward erosion in cancer patients.

Cisplatin (alkylating agents) has been reported to be closely associated with acute coronary thrombosis and even multiple coronary thrombi without any underlying atherosclerosis; it increases the risk of arterial thromboembolic events (ATEs) with and without bleomycin and/or vinca alkaloids (Ito et al. 2012; Karabay et al. 2014). The molecular mechanisms by which cisplatin induce ATEs remain poorly understood and likely involve local and systemic factors. It has been hypothesized that exposure of vascular endothelial cells to chemotherapeutic agents may result in the loss of a thrombo-resistant phenotype and therefore lead to an increased risk of thromboembolic events. Endothelial dysfunction is therefore the key mechanism of altered vasoreactivity. Other mechanisms possibly involved include cisplatin-induced hypomagnesemia, elevated Von Willebrand factor levels, increased platelet aggregation, and the endothelial cell protein C receptor (Hennessy et al. 2002). Cisplatin-induced cytotoxicity in endothelial cells has been linked to increased generation of procoagulant endothelial microparticles and free radicals (Lechner et al. 2007).

Vascular endothelial growth factor (VEGF) signaling pathway inhibitors are another class of drugs associated with the occurrence of angina and other vascular complications such as arterial hypertension. Especially bevacizumab causes ATEs through inhibition of VEGF-A. The incidence of ischemia causing angina or infarction in patients treated with bevacizumab is 1.5%, while the incidence of serious atherothrombotic events is 1.8% (Scappaticci et al. 2007). The ATEs that occur in patients treated with bevacizumab are considered to be due to its anti-vascular endothelial growth factor (VEGF) activity, which decreases the regenerative capability of endothelial cells in response to trauma and further exposes sub-endothelial tissue factors (Ranpura et al. 2010). Endothelial dysfunction plays an important role, as inhibition of VEGF receptor signaling impairs stimulation of endothelial NO synthase (eNOS) activity via the Akt/PKB pathway (Isenberg et al. 2009). Moreover, eNOS uncoupling may occur with an increase in oxidative stress, activation of the endothelin system, furthering the propensity toward abnormal vascular reactivity and structure (Winnik et al. 2013).

Vascular toxicity with bevacizumab can occur at any time during therapy, although in most of the published studies, the median time associated with an event was less than 3 months (Scappaticci et al. 2007).

Other anti-VEGF tyrosine kinase inhibitors such as axitinib, regorafenib, and carbozantinib can cause myocardial infarction, in addition to arterial hypertension, through VEGFR inhibition and endothelial dysfunction with similar mechanism to bevacizumab.

Anti BCR ABL (nilotinib and ponatinib) can cause atherosclerosis progression and acute vascular events such as myocardial infarction, stroke, and peripheral artery occlusive disease (Yang et al. 2015; Iliescu et al. 2016; Di Lisi et al. 2017). Nilotinib inhibits Bcr-Abl, PDGF, cKit, BCR-ABL, PDGFR, c-KIT, CSF-1R, and DDR1. It can accelerate atherosclerosis and peripheral arterial occlusive disease (PAD) and can determine QTc prolongation (Giles et al. 2013). Ponatinib inhibits Bcr-Abl T315I, Src SFKs, and Src e Lyn. It can cause high blood pressure and cardiovascular events (Cortes et al. 2013).

In one of the largest studies regarding the ischemic events triggered by nilotinib, the incidence of all acute vascular events, which included peripheral and cerebral vasospasms, progressive CAD, peripheral artery diseases, and pulmonary embolism was 2% occurring in patients with and without a prior history/diagnosis of atherosclerosis cardiovascular disease or even its risk factors (Quintas-Cardama et al. 2012).

The mechanism that contributes to vascular toxicity of nilotinib and ponatinib will be better explained in the following paragraphs.

2.3 Other Drugs that Cause Myocardial Infarction and Coronary Artery Diseases

Lenalidomide, an immunomodulatory agent used in the management of multiple myeloma, increases the risk for both venous and ATEs, including myocardial infarction and cerebrovascular accidents (Menon et al. 2008).

Long-term hormone deprivation therapies [gonadotropin-releasing hormone (GnRH) agonists] or aromatase inhibitors (anastrozole, letrozole, and exemestane) can promote progression of atherosclerosis and ischemic events by several mechanisms: decreasing insulin sensitivity, altering the plasma lipoprotein particles, and by interfering with the cardio-protective biological effects of testosterone (Nanda et al. 2009; Amir et al. 2011).

Immune checkpoint inhibitors can cause MI; a meta-analysis showed that the incidence of MI in patients treated with these drugs was approximately 1% (Han et al. 2020).

Immune checkpoint inhibitors can contribute to the formation and rupture of unstable atherosclerotic plaques by inducing different inflammatory cytokines and atherosclerotic cytokines related to over-activated T cells (Foks et al. 2017; Tomita et al. 2017).

Also, **anthracyclines** can cause endothelial toxicities attributed to the activation of redox drugs to semiquinone intermediates, which generate superoxide radicals upon reduction (Soultati et al. 2012).

Both the superoxide anion and its dismutation product, hydrogen peroxide, are in fact toxic for the endothelium (Wolf et al. 2006). Endothelial toxicity induced by anthracyclines seems to be influenced by several mechanisms, such as drug accumulation in the nuclei, mitochondria, DNA repair, stress-induced signaling mechanisms, the sarcoplasmic reticulum stress, nitrosative stress, the activity on drug transporters (including MDR1 and MRP1), drug metabolism, and TopI and II inhibition (Menna et al. 2012). The latter is a cellular target for anthracyclines. The ubiquitous TopIIb is expressed in terminally differentiated cells, including adult endothelial cells; hence, it was recently shown that TopIIb could be responsible for the development of anthracycline-induced endothelial toxicity and cardiomyopathy (Capranico et al. 1992). Thus, anthracyclines can also cause negative arterial remodeling.

Radiotherapy can also promote progression of CAD; it causes calcification of the ascending aorta and aortic arch and other vascular lesions inside the radiation field inducing the formation of atherosclerotic lesions (Zhang et al. 2012). The vascular damage induced by irradiation involves the endothelial cells, the ground substance, elastic lamina, smooth muscle cells, and lysosomal activation. Once these elements are impaired, the permeability to circulating lipids increases. ROS has a role in the atherogenic process induced by radiations. In fact, the radiolytic hydrolysis stimulates the production of different types of ROS, including superoxide anion (O^{-2}) (the major product in the presence of oxygen), hydrogen peroxide (H_2O_2), and hydroxyl radical (HO•). Furthermore, radiation promotes inflammation and thrombosis, supporting ROS production, alterations in endothelial cells, and, consequently, vascular damage. It should be noted that together with changes in the endothelial cells, radiation also determines the death of endothelial cells and the resulting exposure of subendothelial thrombotic factors, thereby facilitating vulnerable plaque rupture and thrombotic events (Lancelotti et al. 2013).

The incidence of CAD is linear with the mean cardiac dose of radiotherapy, increasing by 7.4% per Gy; four- to six-fold increase in the risk of CAD in patients received mediastinal radiotherapy. Vascular damage induced by radiotherapy can occur even several years after the termination of exposure to radiations (Giza et al. 2017).

2.4 Arterial Hypertension

Arterial hypertension is a common side effect of VEGF inhibitors with an incidence above 20%. Sunitinib, sorafenib, pazopanib, and other VEGF signaling pathway inhibitors cause arterial hypertension and other vascular side effects.

For example, sunitinib is a multi-target TKI that targets the VEGF receptor (VEGFR) 1–3, PDGFR, c-Kit, FMS-like tyrosine kinase-3 (FLT3), colony-stimulating factor-1 receptor (CSF-1R), and the product of the RET human gene (RET, mutated in medullary thyroid carcinomas/multiple endocrine neoplasia). It can cause high blood pressure and heart failure. Sorafenib is a multi-target TKI that, at clinically relevant concentrations *in vitro* kinase assay, inhibits at least 15 kinases, including VEGFR, PDGFR, Raf-1, B-Raf, c-Kit, and FLT3. It can cause high blood pressure, myocardial ischemia, and, rarely, heart failure. Pazopanib is a small molecule, multi-target inhibitor of PDGFR, VEGFR, and c-KIT. It can cause high blood pressure and congestive heart failure (Di Lisi et al. 2017).

Proposed mechanisms responsible for arterial hypertension include both functional (inactivation of eNOS and production of vasoconstrictors such as endothelin-1) and structural (capillary rarefaction) modifications (Choueiri et al. 2010).

Probably, VEGF-mediated suppression of nephrin, which is important for the maintenance of glomerular function, can contribute to the development of arterial hypertension (Izzedine et al. 2010). The loss of pericytes due to inhibition of PDGFR, along with inhibition of angiogenesis, and due to the VEGFR inhibition is supposed to be the main mechanism for capillary rarefaction (Nazer et al. 2011).

Also the third-generation drug ponatinib is associated with hypertension and ATEs. Endothelial damages leading to pro-atherogenic and anti-angiogenic effects are the suggested mechanisms (Moslehi et al. 2015).

Ibrutinib was also associated with a high incidence of grade 3 or 4 hypertension (Dickerson et al. 2019). The occurrence of arterial hypertension appears to associate with long-term risk for the development of other major cardiac events, including incident arrhythmias.

2.5 Peripheral Artery Occlusive Diseases

Peripheral artery disease (PAD) can occur as complication secondary to anticancer treatment, with an incidence of up to 30% (Zamorano et al. 2016). Especially anti-BCR-ABL (nilotinib and ponatinib) can cause PAD. Nilotinib and ponatinib interact with a considerable number of clinically relevant vascular targets implicated in endothelial cell survival and angiogenesis causing vascular toxicity. The adverse vascular effect of ponatinib was clearly evident in various clinical trials. In five-year follow-up of the phase 2 PACE trial, the cumulative incidence of arterial occlusive events (AOEs), including cardiovascular, cerebrovascular, and peripheral vascular, was 26% (Cortes et al. 2018). These included cardiovascular occlusion in 12% of patients (including coronary artery occlusion and MI, sometimes preceding or concurrent with heart failure), cerebrovascular occlusion (6%), peripheral arterial occlusive events (8%), and venous thromboembolic events (5%) (Cortes et al. 2013; Singh et al. 2020). The precise mechanism of ponatinib induced adverse vascular effects is not clear. Ponatinib can lead to endothelial dysfunction through nonspecific targeting of vascular endothelial growth factor receptors (Gover Proaktor et al. 2017). Furthermore, ponatinib increases the risk of vascular occlusive events by promoting the expression of pro-atherogenic surface adhesion receptors (Valent et al. 2017). Ponatinib also has direct prothrombotic effects by accelerating platelet activation and adhesion (Latifi et al. 2019). In a study, it was demonstrated that in human umbilical endothelial cells, ponatinib induced vascular toxicity through the Notch-1 signaling pathway. The findings from this study revealed that ponatinib inhibits endothelial survival, reduces angiogenesis, and induces endothelial senescence and apoptosis via the Notch-1 pathway. Indeed, selective blockade of Notch-1 prevented ponatinib-induced vascular toxicity (Madonna et al. 2020). Severe cases of nilotinib-associated PAD have recently been reported, occurring in as many as 17% of patients treated (Tefferi et al. 2011). Nilotinib can cause vascular toxicity both directly and indirectly (Quintás-Cardama et al. 2012). In particular, it can promote or exacerbate diabetes mellitus (Mariani et al. 2010), a key risk factor for vascular disease; moreover, it has been shown to produce coronary vasoconstriction in rabbit hearts as well as in isolated human coronary arteries (Lassila et al. 2004). Finally, inhibition of several kinases involved in vascular cell homeostasis such as DDR1, KIT, and/or PDGFR has been invoked as a potential mechanism implicated in nilotinib-induced vascular events (Damrongwatanasuk et al. 2017). However, such kinases are also inhibited by imatinib, an agent that has not been associated with such vascular complications to date (Breccia et al. 2004).

The presence of pre-existing cardiovascular risk factors and pre-existing vascular disease increases the risk of vascular complications during treatment with BCR-ABL inhibitors (Valent et al. 2017). In addition, genetic risk factors may predispose to vascular occlusive disease in patients treated with nilotinib or ponatinib (Bocchia et al. 2016; Zito et al. 2020).

A study showed an increased incidence of AOEs in CML patients treated with ponatinib, with high to very high SCORE risk, suggesting the importance of prophylaxis with 100 mg/day of aspirin and personalized prevention strategies based on CV risk factors (Caocci et al. 2019).

Therefore, patients treated with ponatinib should be informed and closely monitored for high blood pressure and vascular atherothrombotic events.

Although less commonly, acute thrombotic occlusions of the aorta and peripheral arteries can occur with cisplatin therapy (Fernandes et al. 2011). Still, most of the existing literature would point out thrombosis and thromboembolism as the most frequent mechanisms of acute limb ischemia in patients with cancer (Tsang et al. 2011).

2.6 Pulmonary Hypertension

Some anticancer drugs, such as dasatinib, can cause pulmonary arterial hypertension (PAH). Dasatinib inhibits BCR-ABL, SRC family (SRC, LCK, YES, and FYN), c-KIT, EPHA2, and PDGFRβ. It can also cause pleural effusion, heart failure, and pulmonary hypertension (Montani et al. 2012).

The development of PAH with dasatinib increases with prolonged duration of therapy (Jabbour et al. 2014). Experimental studies suggest that this is a consequence of smooth muscle hyperplasia and accompanying endothelial dysfunction (Guignabert et al. 2016). Discontinuation of dasatinib may result in PAH resolution. Few cases of pulmonary arterial hypertension were documented during treatment with nilotinib, bevacizumab, carfilzomib, and rituximab (Lazarevic et al. 2008; McGee et al. 2018).

2.7 Stroke

Stroke has been documented in cancer patients that are at higher risk of thromboembolic events including those related to paradoxical embolization and indwelling catheters (Stefan et al. 2009).

Table 2.1 Vascular events and anticancer drugs.

Vascular events	Anticancer drugs
Myocardial ischemia	VEGF inhibitors, platinum compounds, 5-fluorouracil, gemcitabine, capecitabine, paclitaxel, nilotinib, ponatinib, lenalidomide Long-term hormone deprivation therapies, immune checkpoint inhibitors
Arterial hypertension	VEGF inhibitors, ponatinib
Stroke	VEGF inhibitors, nilotinib, platinum compounds, methotrexate (MTX), 5-fluorouracil, L-asparaginase
Pulmonary hypertension	Dasatinib, nilotinib, bevacizumab, carfilzomib, rituximab
Peripheral artery diseases	Nilotinib, ponatinib

In general, the risk of a chemotherapy-induced stroke is rather low and the risk is higher for some specific regimens such as VEGF inhibitors, nilotinib, platinum compounds, methotrexate (MTX), 5-fluorouracil, and L-asparaginase (Dardiotis et al. 2019); see Table 2.1.

In phase I and II trials of VEGF signaling pathway inhibitors, ischemic stroke and intracranial hemorrhage occurred at a rate of 1.9% each with bevacizumab and in 0% and 3.8% of patients receiving VEGF receptor TKIs, respectively (Fraum et al. 2011).

The underlying causes for the development of a stroke in cancer patients differ from those of non-cancer patients and are associated with the cancer itself as well as with the type of treatment. In general, hypercoagulopathy or other coagulation disorders are most often related to the development of ischemic/embolic stroke. Chemotherapy can lead to stroke via endothelial toxicity and abnormalities in coagulation and hemostasis factors (Saynak et al. 2008).

Chemotherapy releases microparticles from cancer cells, which enhance thrombin generation (Lysov et al. 2017). In addition, radiotherapy can cause vasculopathy through accelerated atherosclerosis or other mechanisms, which can then precipitate stroke (Plummer et al. 2011; Navi and Iadecola 2018).

Cisplatin has repeatedly been reported to be associated with cardiovascular events. However, the mechanisms involved remain largely unknown. Circulating endothelial- and platelet-derived particles can contribute to cisplatin-induced stroke (Grisold et al. 2009).

2.8 Thromboembolic Risk and Atrial Fibrillation

Atrial fibrillation (AF) contributes to increase the risk of stroke. Patients with cancer exhibited a 20% higher adjusted risk of atrial fibrillation compared to those without cancer (O'Neal et al. 2015).

In cancer patients, new onset AF is associated with a 2-fold higher risk of thromboembolism complications and a 6-fold higher risk of heart failure as well as a 10-fold higher risk of 30-day mortality, even after adjusting for known risk factors (Rahman et al. 2016; Alexandre et al. 2020).

Especially ibrutinib, abiraterone, anthracyclines, gemcitabine, paclitaxel, docetaxel, dacarbazine, cisplatin, rituximab, interleukin-2, bortezomib, and ipilimumab can contribute to the onset of atrial fibrillation through multiple mechanisms (Alexandre et al. 2018).

The most common underlying factors include release of pro-inflammatory cytokines, abnormalities in calcium homeostasis, and direct myocardial damage. In addition, the increase in vagal and adrenergic tones, often due to hypotension, myocardial ischemia, and abnormal electrolyte concentrations, may be involved, especially when using alkylating agents, anthracyclines, antimetabolites, docetaxel, 5-fluorouracil, gemcitabine, rituximab, paclitaxel, alemtuzumab, and etanercept (Suter et al. 2013). Additional arrhythmogenic substrates include coronary vasospasm, through inhibition of endothelial nitric oxide synthesis and generation of reactive oxygen species mediating oxidative damage on the vessel wall, and a direct cardiotoxic effect on the atrial conduction system. All these effects are combined with a systemic pro-inflammatory state, which is typical of malignancies (Hu et al. 2015).

Ibrutinib, an oral Bruton's tyrosine kinase inhibitor, is associated with high incidence of atrial fibrillation and cardiovascular toxicity. It targets several alternative kinases, broadening its efficacy as an effective immune modulator (Byrd et al. 2013).

In a study, the risk of developing AF with ibrutinib was 15-fold higher than that expected in the general population and in the population of patients with cancer not exposed to ibrutinib (Baptiste et al. 2019). The pathogenesis of atrial fibrillation induced by ibrutinib remains poorly understood; probably, the association between common AF risk factors and the direct effects of ibrutinib treatment contributes to AF onset. The CHA2DS2-VASc and HAS-BLED scores, although not validated in cancer patients, are usually applied to determine individual thrombotic and bleeding risks, respectively (Tufano et al. 2018). The use of anticoagulant agents during cancer therapy poses a particular challenge because of the risks, drug–drug interactions and bleeding. Bleeding risk appeared not to be equivalent among the type of malignancy, with gastrointestinal cancer being more at risk.

2.9 Prevention and Treatment of Vascular Toxicity

There are no specific guidelines for the treatment of vascular toxicity in cancer patients, but roughly those used in the general population are also used in cancer patients. Since cancer and some anticancer-drugs contribute to increase the thromboembolic risk and vascular toxicity, cardiovascular prevention and careful follow-up are needed in cancer patients (Lyon et al. 2020). Thus, baseline cardiovascular risk correction is recommended in all patients scheduled to receive drugs with potential cardiotoxicity to reduce the risk of developing complications. Signs and symptoms of vascular damage should be evaluated and in the suspicion of CAD, PAD, or cerebrovascular disease, further test should be performed in order to better stratify patients risk and to optimize medical therapy. In high-risk patients treated with ponatinib, preventive treatment with aspirin should be considered (Lyin et al. 2020).

Certainly, a close cooperation between oncologist and cardiologist is needed to prevent and treat vascular toxicity.

References

1. Alexandre J, Molsehi JJ, Bersell KR, et al. Anticancer drug induced cardiac rhythm disorders: Current knowledge and basic underlying mechanisms. Pharmacol Ther. 2018;189:89–103.
2. Alexandre J, Salem J, Moslehi J, et al. Identification of anticancer drugs associated with atrial fibrillation - analysis of the WHO pharmacovigilance database. Eur Heart J Cardiovasc Pharmacother. 2020;pvaa037.
3. Amir E, Seruga B, Niraula S, et al. Toxicity of adjuvant endocrine therapy in postmeno- pausal breast cancer patients: a systematic review and meta-analysis. J Natl Cancer Inst. 2011;103:1299– 309.
4. Anand AJ. Fluorouracil cardiotoxicity. Ann Pharmacother. 1994;28:374–8.
5. Arima Y, Oshima S, Noda K, et al. Sorafenib-induced acute myocardial infarction due to coronary artery spasm. J Cardiol. 2009;54:512–515.
6. Armitage JD, Montero C, Benner A, et al. Acute coronary syndromes complicating the first infusion of rituximab. Clin Lymphoma Myeloma. 2008;8:253–255.
7. Baptiste F, Cautela J, Ancedy Y, et al. High incidence of atrial fibrillation in patients treated with ibrutinib.Open Heart. 2019;6:e001049.
8. Bocchia M, Galimberti S, Aprile L, et al. Genetic predisposition and induced pro-inflammatory/pro-oxidative status may play a role in increased atherothrombotic events in nilotinib treated chronic myeloid leukemia patients. Oncotarget. 2016;7:72311–21.
9. Breccia M, Muscaritoli M, Aversa Z, et al. Imatinib mesylate may improve fasting blood glucose in diabetic Ph + chronic myelogenous leukemia patients responsive to treatment. J Clin Oncol.2004;22:4653–5.

10. Byrd JC, Furman RR, Coutre SE, et al. Targeting BTK with ibrutinib in relapsed chronic lymphocytic leukemia. N Engl J Med. 2013;369:32-42.
11. Caocci G, Mulas O, Abruzzese E, et al. Arterial occlusive events in chronic myeloid leukemia patients treated with ponatinib in the real-life practice are predicted by the Systematic Coronary Risk Evaluation (SCORE) chart. Hematological Oncology. 2019;37:296–302.
12. Capranico G, Tinelli S, Austin CA, et al. Different patterns of gene expression of topoisomerase II isoforms in differentiated tissues during murine development. Biochim Biophys Acta. 1992;1132:43-8.
13. Cardinale D, Colombo A, Colombo N. Acute coronary syndrome induced by oral capecitabine. Can J Cardiol. 2006;22:251–253.
14. Chong JH, Ghosh AK. Coronary Artery Vasospasm Induced by 5-fluorouracil: Proposed Mechanisms, Existing Management Options and Future Directions. Interv Cardiol. 2019; 14: 89–94.
15. Choueiri TK, Schutz FAB, Je Y, et al. Risk of arterial thromboembolic events with sunitinib and sorafenib: a systematic review and meta-analysis of clinical trials. J Clin Oncol. 2010;28:2280–5.
16. Cortes JE, Kim DW, Pinilla Ibarz J, et al. A phase 2 trial of ponatinib in Philadelphia chromosome positive leukemias. N Engl J Med. 2013;369:1783-96.
17. Cortes JE, Kim DW, Pinilla Ibarz J, et al. Ponatinib efficacy and safety in Philadelphia chromosome positive leukemia: final 5 year results of the phase 2 PACE trial. Blood. 2018;132:393 -404.
18. Cortes JE, Kim DW, Pinilla-Ibarz J, et al; PACE Investigators. A phase 2 trial of ponatinib in Philadelphia chromosome-positive leukemias. N Engl J Med. 2013;369:1783-96.
19. Damrongwatanasuk R, Fradley MG. Cardiovascular Complications of Targeted Therapies for Chronic Myeloid Leukemia. Curr Treat Options Cardiovasc Med. 2017;19:24.
20. Dardiotis E, Aloizou AM, Markoula S, et al. Cancer-associated stroke: Pathophysiology, detection and management. Int J Oncol. 2019;54:779-796.
21. Davignon J GP. Role of endothelial dysfunction in atherosclerosis. Circulation 2004;109(Suppl 1):11127–32.
22. Di Lisi D, Madonna R, Zito C, et al. Anticancer therapy-induced vascular toxicity: VEGF inhibition and beyond. Int J Cardiol. 2017; 227:11-17.
23. Dickerson T, Wiczer T, Waller A, et al. Hypertension and incident cardiovascular events following ibrutinib initiation. Blood. 2019;134:1919-1928.

24. Escudier B, Eisen T, Stadler WM, et al. Sorafenib in advanced clear-cell renal- cell carcinoma. N Engl J Med. 2007;356:125–34.
25. Fernandes DD, Louzada ML, Souza CA, et al. Acute aortic thrombosis in patients receiving cisplatin-based chemotherapy. Curr Oncol. 2011;18:e97–e100.
26. Foks AC, Kuiper J. Immune checkpoint proteins: exploring their therapeutic potential to regulate atherosclerosis. Br J Pharmacol. 2017;174:3940e3955.
27. Fraum TJ, Kreisl TN, Sul J, et al. Ischemic stroke and intracranial hemorrhage in glioma patients on antiangiogenic therapy. J Neurooncol. 2011;105:281–289.
28. Giles FJ, Mauro MJ, Hong F, et al. Rates of peripheral arterial occlusive disease in patients with chronic myeloid leukemia in the chronic phase treated with imatinib, nilotinib, or non-tyrosine kinase therapy: a retrospective cohort analysis. Leukemia. 2013;27:1310-5.
29. Giza DE, Boccalandro F, Lopez-Mattei J, et al. Ischemic heart disease: special considerations in cardio-oncology. Curr Treat Options Cardiovasc Med 2017;19:37.
30. Giza DE, Boccalandro F, Lopez-Mattei J, et al. Ischemic Heart Disease: Special Considerations in Cardio-Oncology. Curr Treat Options Cardiovasc Med. 2017;19:37.
31. Gover Proaktor A, Granot G, Shapira S, et al. Ponatinib reduces viability, migration, and functionality of human endothelial cells. Leuk Lymphoma. 2017;58:1455-67.
32. Grisold W, Oberndorfer S, Struhal W. Stroke and cancer: A review. Acta Neurol Scand. 2009;119:1–16.
33. Guignabert C, Phan C, Seferian A, et al. Dasatinib induces lung vascular toxicity and predisposes to pulmonary hypertension. J Clin Invest. 2016;126:3207–3218.
34. Han XJ, Li JQ, Khannanova Z, et al. Optimal management of coronary artery disease in cancer patients. Chronic Dis Transl Med. 2020;5:221-233.
35. Hennessy B, O'Connor M, Carney DN. Acute vascular events associated with cisplatin therapy in malignant disease. Ir Med J 2002;95:145–6, 148.
36. Herrmann J, Yang EH, Iliescu CA, et al. Vascular Toxicities of Cancer Therapies: The Old and the New--An Evolving Avenue. Circulation. 2016;133:1272-89.
37. Hu YF, Chen YJ, Lin YJ, et al. Inflammation and the pathogenesis of atrial fibrillation. Nat Rev Cardiol 2015;12: 230–243

38. Iliescu CA, Grines CL, Herrmann J, et al. SCAI Expert consensus statement: Evaluation, management, and special considerations of cardio-oncology patients in the cardiac catheterization laboratory (endorsed by the cardiological society of india, and sociedad Latino Americana de Cardiologıa intervencionista). Catheter Cardiovasc Interv. 2016;87:E202-23.
39. Isenberg JS, Martin-Manso G, Maxhimer JB, et al. Regulation of nitric oxide signalling by thrombospondin 1: Implications for anti-angiogenic therapies. Nat Rev Cancer. 2009;9:182–194.
40. Ito D, Shiraishi J, Nakamura T, et al. Primary percutaneous coronary intervention and intravascular ultrasound imaging for coronary thrombosis after cisplatin-based chemotherapy. Heart Vessels. 2012;27:634–638.
41. Izzedine H, Massard C, Spano JP, et al. VEGF signalling inhibition-induced proteinuria: Mechanisms, significance and management. Eur J Cancer. 2010;46:439-48.
42. Jabbour E, Kantarjian HM, Saglio G, et al. Early response with dasatinib or imatinib in chronic myeloid leukemia: 3-year follow-up from a randomized phase 3 trial (DASISION). Blood. 2014; 123:494–500.
43. Karabay KO, Yildiz O, Aytekin V. Multiple coronary thrombi with cisplatin. J Invasive Cardiol. 2014; 26:E18–20.
44. Kosmas C, Kallistratos MS, Kopterides P, et al. Cardiotoxicity of fluoropyrimidines in different schedules of administration: a prospective study. J Cancer Res Clin Oncol. 2008;134:75–82.
45. Lancellotti P, Nkomo VT, Badano LP, et al. ; European Society of Cardiology Working Groups on Nuclear Cardiology and Cardiac Computed Tomography and Cardiovascular Magnetic Resonance; American Society of Nuclear Cardiology; Society for Cardiovascular Magnetic Resonance; Society of Cardiovascular Computed Tomography. Expert consensus for multi-modality imaging evaluation of cardiovascular complications of radiotherapy in adults: a report from the European Association of Cardiovascular Imaging and the American Society of Echocardiography. Eur Heart J Cardiovasc Imaging. 2013;14:721-40. Erratum in: Eur Heart J Cardiovasc Imaging. 2013;14:1217.
46. Lassila M, Allen TJ, Cao Z, et al. Imatinib attenuates diabetesassociated atherosclerosis. Arterioscler Thromb Vasc Biol. 2004;24:935–42.
47. Latifi Y, Moccetti F, Wu M, et al. Thrombotic microangiopathy as a cause of cardiovascular toxicity from the BCR-ABL1 tyrosine kinase inhibitor ponatinib. Blood. 2019;133:1597-606.

48. Lazarevic V, Liljeholm M, Forsberg K, et al. Fludarabine, cyclophosphamide and rituximab (FCR) induced pulmonary hypertension in Waldenstrom macroglobulinemia. Leuk Lymphoma. 2008;49: 1209–1211.
49. Lechner D, Kollars M, Gleiss A, et al. Chemotherapy-induced thrombin generation via procoagulant endothelial microparticles is independent of tissue factor activity. J Thromb Haemost. 2007;5:2445-52.
50. Lu JI, Carhart RL, Graziano SL, et al. Acute coronary syndrome secondary to fluorouracil infusion. J Clin Oncol. 2006;24:2959–2960.
51. Lyon AR, Dent S, Stanway S, et al. Baseline cardiovascular risk assessment in cancer patients scheduled to receive cardiotoxic cancer therapies: a position statement and new risk assessment tools from the Cardio-Oncology Study Group of the Heart Failure Association of the European Society of Cardiology in collaboration with the International Cardio-Oncology Society. Eur J Heart Fail. 2020;22:1945-1960.
52. Lysov Z, Dwivedi DJ, Gould TJ, et al. Procoagulant effects of lung cancer chemotherapy: Impact on microparticles and cell-free DNA. Blood Coagul Fibrinolysis. 2017;28:72–82.
53. Madonna R, Pieragostino D, Cufaro MC, et al. Ponatinib Induces Vascular Toxicity through the Notch-1 Signaling Pathway. J Clin Med. 2020;9.
54. Mariani S, Tornaghi L, Sassone M, et al. Imatinib does not substantially modify the glycemic profile in patients with chronic myeloid leukaemia. Leuk Res. 2010;34:e5–7.
55. McGee M, Whitehead N, Martin J, et al. Drug-associated pulmonary arterial hypertension. Clin Toxicol (Phila) 2018;56:801-809.
56. Menna P, Paz OG, Chello M, et al. Anthracycline cardiotoxicity. Expert Opin Drug Saf. 2012;11 S1:S21-36.
57. Menon SP, Rajkumar SV, Lacy M, et al. Thromboembolic events with lenalidomide-based therapy for multiple myeloma. Cancer. 2008;112:1522–8.
58. Montani D, Bergot E, Günther S, et al. Pulmonary arterial hypertension in patients treated by dasatinib. Circulation. 2012;125:2128-37.
59. Moslehi JJ, Deininger M. Tyrosine kinase inhibitor-associated cardiovascular toxicity in chronic myeloid leukemia. J Clin Oncol. 2015;10:4210–8.
60. Nanda A, Chen MH, Braccioforte MH, et al. Hormonal therapy use for prostate cancer and mortality in men with coronary artery disease-induced congestive heart failure or myocardial infarction. JAMA. 2009;302:866-73.

61. Navi BB, Iadecola C. Ischemic stroke in cancer patients: A review of an underappreciated pathology. Ann Neurol. 2018;83:873-883.
62. Nazer B, Humphreys BD, Moslehi J. Effects of novel angiogenesis inhibitors for the treatment of cancer on the cardiovascular system: focus on hypertension. Circulation. 2011;124:1687-91.
63. O'Neal WT, Lakoski SG, Qureshi W, et al. Relation between cancer and atrial fibrillation (from the REasons for Geographic And Racial Differences in Stroke Study). Am J Cardiol 2015;115:1090–1094.
64. Ozturk B, Tacoy G, Coskun U, et al. Gemcitabine-induced acute coronary syndrome: A case report. Med Princ Pract. 2009;18:76–80.
65. Plummer C, Henderson RD, O'Sullivan JD, et al. Ischemic stroke and transient ischemic attack after head and neck radiotherapy: A review. Stroke. 2011;42:2410–2418.
66. Quintás-Cardama A, Kantarjian H, Cortes J. Nilotinib-associated vascular events. Clin Lymphoma Myeloma Leuk. 2012;12:337-40.
67. Quintas-Cardama A, Kantarjian H, Cortes J. Nilotinibassociated vascular events. Clinical lymphoma, myeloma & leukemia. 2012;12:337–40.
68. Rahman F, Ko D, Benjamin EJ. Association of Atrial Fibrillation and Cancer. JAMA Cardiol 2016;1:384–386.
69. Ranpura V, Hapani S, Chuang J, et al. Risk of cardiac ischemia and arterial thromboembolic events with the angiogenesis inhibitor bevacizumab in cancer patients: a meta-analysis of randomized controlled trials. Acta Oncol. 2010;49:287–97.
70. Razak NB, Jones G, Bhandari M, et al. Cancer-Associated Thrombosis: An Overview of Mechanisms, Risk Factors, and Treatment. Cancers. 2018; 10: 380.
71. Saynak M, Cosar-Alas R, Yurut-Caloglu V, et al. Chemotherapy and cerebrovascular disease. J BUON. 2008;13:31–36.
72. Scappaticci FA, Skillings JR, Holden SN, et al. Arterial thromboembolic events in patients with metastatic carcinoma treated with chemotherapy and bevacizumab. J Natl Cancer Inst. 2007;99:1232–9.
73. Schrader C, Keussen C, Bewig B, et al. Symptoms and signs of an acute myocardial ischemia caused by chemotherapy with paclitaxel (taxol) in a patient with metastatic ovarian carcinoma. Eur J Med Res. 2005;10:498–501.
74. Senkus E, Jassem Jl. Cardiovascular effects of systemic cancer treatment. Cancer Treat Rev. 2011;37:300–11.
75. Singh AP, Umbarkar P, Tousif S, et al. Cardiotoxicity of the BCR-ABL1 tyrosine kinase inhibitors: Emphasis on ponatinib. Int J Cardiol. 2020;316:214-221.

76. Soultati A, Mountzios G, Avgerinou C, et al. Endothelial vascular toxicity from chemotherapeutic agents: preclinical evidence and clinical implications. Cancer Treat Rev. 2012;38:473-83.
77. Stefan O, Vera N, Otto B, et al. Stroke in cancer patients: A risk factor analysis. J Neurooncol. 2009;94:221–226.
78. Suter TM, Ewer MS. Cancer drugs and the heart: importance and management. Eur Heart J. 2013;34:1102–1111.
79. Tefferi A, Letendre L. Nilotinib treatment-associated peripheral artery disease and sudden death: yet another reason to stick to imatinib as front-line therapy for chronic myelogenous leukemia. Am J Hematol 2011; 86:610-1.
80. Thyss A, Gaspard MH, Marsault R et al. Very high endothelin plasma levels in patients with 5-FU cardiotoxicity. Ann Oncol. 1992;3:88.
81. Tomita Y, Sueta D, Kakiuchi Y, et al. Acute coronary syndrome as a possible immune-related adverse event in a lung cancer patient achieving a complete response to anti-PD-1 immune checkpoint antibody. Ann Oncol. 2017;28:2893e2895.
82. Tsang JS, Naughton PA, O'Donnell J, et al. Acute limb ischemia in cancer patients: Should we surgically intervene? Ann Vasc Surg. 2011; 25:954–960.
83. Tufano A, Galderisi M, Esposito L, et al. Anticancer Drug-Related Nonvalvular Atrial Fibrillation: Challenges in Management and Antithrombotic Strategies. Semin Thromb Hemost 2018;44:388–396.
84. Valent P, Hadzijusufovic E, Hoermann G, et al. Risk factors and mechanisms contributing to TKI-induced vascular events in patients with CML. Leuk Res. 2017;59:47–54.
85. Van Cutsem E, Hoff PM, Blum JL, et al. Incidence of cardiotoxicity with the oral fluoropyrimidine capecitabine is typical of that reported with 5-fluorouracil. Ann Oncol 2002;13:484–5.
86. Wilhelm SM, Adnane L, Newell P, et al. Preclinical overview of sorafenib, a multikinase inhibitor that targets both Raf and VEGF and PDGF receptor tyrosine kinase signaling. Mol Cancer Ther. 2008;7:3129-40.
87. Winnik S, Lohmann C, Siciliani G, et al. Systemic vegf inhibition accelerates experimental atherosclerosis and disrupts endothelial homeostasis--implications for cardiovascular safety. Int J Cardiol. 2013;168:2453–2461.
88. Wolf MB, Baynes JW. The anti-cancer drug, doxorubicin, causes oxidant stress-induced endothelial dysfunction. Biochim Biophys Acta. 2006;1760:267-71.

89. Yang EH, Watson KE, Herrmann J. Should vascular effects of newer treatments be addressed more completely? Future Oncol. 2015;11:1995-8.
90. Yasue H, Nakagawa H, Itoh T, et al. Coronary artery spasm--clinical features, diagnosis, pathogenesis, and treatment. J Cardiol. 2008;51:2-17.
91. Zamorano JL, Lancellotti P, Rodriguez Muñoz D, et al. 2016 ESC position paper on cancer treatments and cardiovascular toxicity developed under the auspices of the ESC committee for practice guidelines: The task force for cancer treatments and cardiovascular toxicity of the European society of cardiology (ESC). Eur Heart J. 2016;37:2768–801.
92. Zhang S, Liu X, Bawa-Khalfe T, etal. Identification of the molecular basis of doxorubicin-induced cardiotoxicity. Nat Med. 2012;18:1639-42
93. Zito C, Manganaro R, Carerj S, et al. Peripheral Artery Disease and Stroke. J Cardiovasc Echogr. 2020; 30(S1): S17–S25.

3

Hypertensive Oncologic Patients

Giacomo Tini[1], Massimo Volpe[1], Paolo Spallarossa[2]

[1]Cardiology, Azienda Ospedaliero-Universitaria Sant'Andrea, University of Rome Sapienza, Rome, Italy
[2]Clinic of Cardiovascular Diseases, IRCCS Ospedale Policlinico San Martino, Genova, Italy

Correspondence to:
Paolo Spallarossa, MD, Clinic of Cardiovascular Diseases, IRCCS Ospedale Policlinico San Martino, Largo Rosanna Benzi, 10 16132, Genova, Italy
paolo.spallarossa@unige.it

KEYWORDS: Anthracycline; Anti-VEGF; Arterial Hypertension; Cardiotoxicity; Cardiovascular Prevention; Cardiovascular Risk Factors.

3.1 Introduction

Approximately one out of four adults in the world is affected by arterial hypertension, the most common risk factor for cardiovascular (CV) and cerebrovascular diseases, which are the main contributors to mortality and morbidity worldwide (Williams et al. 2018).

Prevalence of hypertension in cancer patients is similar to that of general population; hence, it is the most common comorbidity in this condition (Piccirillo et al. 2004; Unger et al. 2019).

The importance of hypertension in the oncologic patient is related to the increased susceptibility it confers toward cardiotoxicity due to anticancer treatments (Tini et al. 2019). More in general, major CV risk factors predispose to cardiotoxicity (Pinder et al. 2007; Cameron et al. 2016; Tini et al. 2019) and a worse CV risk profile has been associated with cardiac events and adverse outcomes during and after oncologic treatments (Armenian et al. 2016; Hershman et al. 2018).

It thus appears clear that a hypertensive oncologic patient carries a higher risk of CV events in the short and long terms and requires a particularly tailored management.

Chapter III will highlight the main issues related to the co-existence of arterial hypertension and cancer.

3.2 Implications of Arterial Hypertension in Patients Undergoing Anticancer Treatments

Occurrence of cardiotoxicity mainly depends on two factors: the type of anticancer treatment with its inherent toxicity and the individual CV risk profile (Cameron et al. 2016). In this regard, arterial hypertension is typically considered the most important CV factor favoring cardiotoxicity (Tini et al. 2019).

The mechanisms through which arterial hypertension enhances the risk of cardiotoxicity are multiple and depend on the type of anticancer treatment. Here follow three examples.

3.2.1 Anthracyclines

In the case of anthracyclines, arterial hypertension acts as an *add-on* stressor together with the direct myocardial damage caused by the anticancer drug (Spallarossa et al. 2016; Tini et al., 2019).

Cardiotoxicity induced by anthracyclines occur with a characteristic timing, as proposed by the "multiple-stress" or "multiple-hit" hypothesis (Spallarossa et al. 2016). In brief, the damage on cardiomyocytes caused by anthracyclines may represent the substrate for a second, subsequent stressor (which may be a long-standing hypertension as well as merely cardiac aging), or may itself be the second stressor, acting upon an already damaged heart, as in the case of hypertensive heart disease (Hahn et al. 2014). Hypertension has been indeed identified as one of the strongest predictors of heart failure after anthracycline therapy in breast cancer women older than 65 years (Pinder et al. 2007). Moreover, patients with pre-existing hypertension treated with anthracycline were also shown to be significantly more likely to undergo a discontinuation of the anticancer therapy or a delay in treatment or a reduction in anthracycline doses, implying that hypertension may also render less effective chemotherapy, with important prognostic implications (Szmit et al. 2014).

3.2.2 Carfilzomib

Proteasome inhibitors, mainstay of therapy for multiple myeloma, have been related to the risk of cardiotoxicity, in particular in patients at high CV risk

(Bringhen et al. 2018; Lee et al. 2018). This is particularly true for carfilzomib, a second-generation agent. In trials assessing its efficacy, incidence of new-onset hypertension was up to 25% and new heart failure was 10%. Mechanisms of carfilzomib cardiotoxicity are still not fully elucidated. However, data from follow-up of phase 3 trials on carfilzomib seem to show that the long-term risk of heart failure is not augmented (Chari et al. 2018). Carfilzomib induces a rise in blood pressure through endothelial dysfunction and dysregulation of nitric oxide (NO) production (Chari et al. 2014). It has been proposed that the risk of heart failure with carfilzomib is related to the transient increase in blood pressure, paired with significant hydration and concurrent steroid therapy that multiple myeloma patients usually receive. Moreover, these patients are often elderly and present a high CV risk (Bringhen et al. 2018; Chari et al. 2018). In this context, it has been shown that among recipients of carfilzomib, those at higher risk of developing CV adverse events were patients with a pre-existing arterial hypertension mediated organ damage, such as left ventricular hypertrophy (Bruno et al. 2008).

3.2.3 Anti-vascular endothelial growth factor agents

Among anticancer treatments, anti-vascular endothelial growth factor (VEGF) agents are those most commonly associated with the risk of developing new-onset hypertension (Tini et al. 2019). In a large meta-analysis of 77 studies investigating different anti-VEGF agents (both monoclonal antibodies directed toward VEGF and tyrosine kinase inhibitors), hypertension was the most common side effect, with a number needed to harm of 6 (Abdel et al. 2017).

Anti-VEGF agents cause an increase in blood pressure through several pathways. First, reduction in NO production due to inhibition of VEGF receptor-2 determines vasoconstriction and augmented peripheral resistances. Moreover, VEGF receptors are also expressed in the kidney and their inhibition may induce glomerular lesion and, ultimately, proteinuria and/or worsening renal function. Anti-VEGF agents also increase levels of endothelin 1, a molecule with vasocostrictive effect. Furthermore, VEGF inhibition may cause endothelial cell apoptosis, resulting in a phenomenon called microcapillary rarefactions, and induce renal thrombotic microangiopathy (Russo et al. 2019). Finally, it has been shown that anti-VEGF agents determine a direct myocardial damage by overproduction of reactive oxygen species (Neves et al. 2018).

Since virtually all patients treated with anti-VEGF agents present an increase in blood pressure, it is clear how individuals with an already known arterial hypertension are at higher risk of reaching pathologic values. Though blood pressure increase due to anti-VEGF agents may be rapid and even severe, this adverse effect appears to be easy to manage if promptly treated (Tini et al. 2019). As in the case of carfilzomib, the risk of heart failure due to anti-VEGF agents is likely to be, at least in part, related to hypertensive disorders.

3.3 Management of Arterial Hypertension in Oncologic Patients

Despite being recognized as the number one comorbidity of cancer patients and a compelling issue related to the risk of cardiotoxicity, arterial hypertension management is often overlooked in the oncologic setting.

However, recently, the topic of arterial hypertension in cancer patients has been addressed also in the latest European Society of Cardiology guidelines (Williams et al. 2018), and, more in general, a greatest attention toward CV prevention in oncology has been paid (Yin et al. 2019; Tini et al. 2019).

To pursue an optimal management of arterial hypertension in the oncologic patient, a step-up approach dedicated to the whole CV profile evaluation is needed.

1. *Baseline CV assessment:* The importance of a baseline CV assessment, i.e., prior to initiation of an anticancer treatment, has been recognized (Spallarossa et al. 2016; Lyon et al. 2020). This is important for several reasons. First, CV risk factors often cluster in the same patient; thus, one cannot look only at arterial hypertension without searching for other risk factors (dyslipidemia, diabetes mellitus, obesity, etc.) and assessing the overall CV profile. In fact, blood pressure targets are linked to the overall CV risk profile. Second, in the oncologic setting, CV risk factors are usually defined based on the clinical history (i.e., present vs. absent), regardless of whether they are controlled or not (Tini et al. 2020). On the contrary, the influence of a given risk factor may be dramatically different if it is at target or not or if its organ damage is already detectable (as in the case of arterial hypertension-mediated organ damage) (Tini et al. 2019). Importantly, the baseline CV assessment of an oncologic patient should not mandatorily be performed by a cardio-oncologist or a cardiologist. Not only any clinical practitioner – as the same referring oncologist – may perform a basic CV evaluation, but this may also help to differentiate those individuals requiring a specialistic cardiologic consultation from those at low CV risk, not needing further, possibly time-consuming, evaluations. For example, patients may

present an already-known (even well-controlled) arterial hypertension or may present no CV risk factors. But "grey" cases may occur, as with individuals with previously unknown arterial hypertension or with normal-high blood pressure values and overall high-risk CV profile which requires optimization.

2. *Consider the oncologic setting:* After having assessed and optimized the overall CV profile, it may be helpful to consider the oncologic setting and the anticancer treatment the patient is scheduled for. Few are the situations in which a specific baseline CV therapy may be required. For example, if a patient scheduled to receive anthracyclines presents a high-risk CV profile or uncontrolled arterial hypertension, it is reasonable to introduce (or up-titrate) cardio-active drugs, such as second-generation beta-blockers and angiotensin converting enzyme-inhibitors (ACEi) or angiotensin receptor blockers (ARBs).
   In the case of drugs which may cause new-onset or worsen an already existing hypertension (as anti-VEGF agents), it may be useful to *plan* an anti-hypertensive treatment. This means that the patient should be educated regarding the risk of increasing blood pressure values and, thus, to check them frequently. In our everyday practice, we suggest to the referring oncologist and to the patient a starting treatment, and one or two up-titration steps, to allow a prompt initiation of anti-hypertension management and to avoid delay in anticancer treatment delivery. We usually suggest an initial treatment with ACEi/ARBs and dihydropyridine calcium channel blockers. We further believe that in the case of hypertension due to anti-VEGF agents, anti-hypertensive medicaments targeting NO should be used cautiously. Since NO and its pathway are involved in angiogenesis and are on-target of these drugs, there are no data that addressing NO would not compromise the anti-angiogenetic, anti-cancer effect of anti-VEGF agents (Russo et al. 2019). Such an approach, with prompt management and stringent blood pressure targets, reflects the current advices by the European Society of Cardiology, suggested for the overall population (Williams et al. 2018). A practical algorithm is provided in Figure 3.1.
3. *Do not forget the patient status:* As for hypertension, also arterial hypotension is often overlooked in cancer patients. Cardiac drugs may have side effects and cause changes in sympathetic tone, volemia, heart rate, etc. Cancer itself and antineoplastic treatments are associated with fatigue and hemodynamic alterations. Cardioactive therapies may thus concur to debilitate the patient (Sarocchi et al. 2018). The CV evaluation must also deal with this issue. The need for suspending or reducing a

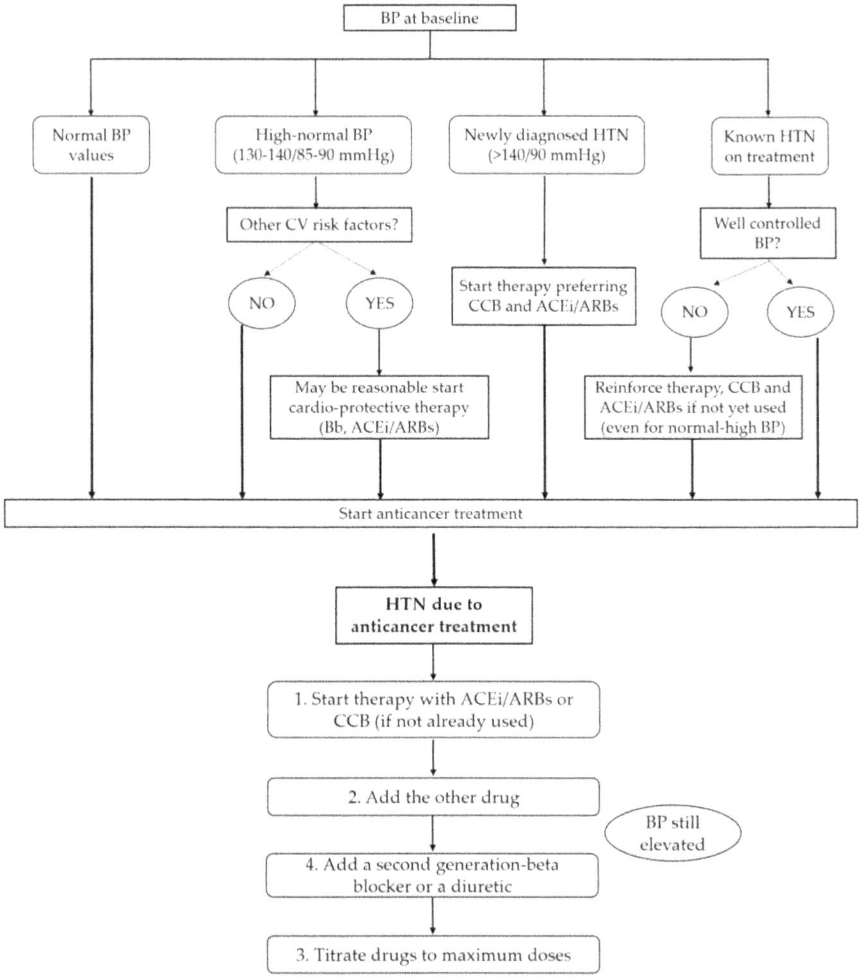

Figure 3.1 Algorithm for evaluation of blood pressure and management of arterial hypertension in cancer patients [modified from Spallarossa et al. (Russo et al. 2019)]. (BP: blood pressure, HTN: hypertension, Bb: beta blockers, CCB: calcium channel blockers, ACEi: angiotensin-converting enzyme inhibitors, ARBs: angiotensin receptor blockers)

previous therapy for hypertension in a cancer patient is not infrequent. Patients should be educated to know how to act with anti-hypertension therapy when blood pressure is low. A similar, inverse strategy to that proposed above may be considered.

4. *Remember the importance of communication with the patient:* The baseline CV assessment may be the chance to improve CV health education and awareness in the oncologic setting (Battistoni et al. 2015, 2019). Cancer

patients may perceive any CV issue as a minor problem as compared to the oncologic one. It is the duty of the physician to explain that not caring for CV health can jeopardize the success of anticancer treatments. It has been shown that non-adherence to CV medications after an anticancer treatment is associated with the risk of cardiac events and adverse outcomes (Hershman et al. 2020). Thus, the mission of cardio-oncology should also be of increasing CV awareness among cancer patients.

References

1. Abdel-Qadir H, Ethier J-L, Lee DS, et al. Cardiovascular toxicity of angiogenesis inhibitors in treatment of malignancy: A systematic review and meta-analysis. Cancer Treat Rev. 2017;53:120–7.
2. Armenian SH, Xu L, Ky B, et al. Cardiovascular Disease Among Survivors of Adult-Onset Cancer: A Community-Based Retrospective Cohort Study. J Clin Oncol. 2016;34:1122–30.
3. Battistoni A, Mastromarino V, Volpe M. Reducing Cardiovascular and Cancer Risk: How to Address Global Primary Prevention in Clinical Practice. Clin Cardiol. 2015;38:387–94.
4. Battistoni A, Tocci G, Presta V, et al.. Antihypertensive drugs and the risks of cancer: More fakes than facts. Eur J Prev Cardiol. 2019;2047487319884823.
5. Bringhen S, Milan A, Ferri C, et al. Cardiovascular adverse events in modern myeloma therapy – Incidence and risks. A review from the European Myeloma Network (EMN) and Italian Society of Arterial Hypertension (SIIA). Haematologica. 2018;103:1422–32.
6. Bruno G, Bringhen S, Maffei I, al. Cardiovascular Organ Damage and Blood Pressure Levels Predict Adverse Events in Multiple Myeloma Patients Undergoing Carfilzomib Therapy. Cancers. 2019;11:622.
7. Cameron AC, Touyz RM, Lang NN. Vascular Complications of Cancer Chemotherapy. Can J Cardiol. 2016;32:852–62.
8. Chari A, Hajje D. Case series discussion of cardiac and vascular events following carfilzomib treatment: possible mechanism, screening, and monitoring. BMC Cancer. 2014;14.
9. Chari A, Stewart AK, Russell SD, et al. Analysis of carfilzomib cardiovascular safety profile across relapsed and/or refractory multiple myeloma clinical trials. Blood Adv. 2018;2:1633–44.
10. Hahn VS, Lenihan DJ, Ky B. Cancer Therapy–Induced Cardiotoxicity: Basic Mechanisms and Potential Cardioprotective Therapies. J Am Heart Assoc. 2014;3(2).

11. Hershman DL, Accordino MK, Shen S, et al. Association between nonadherence to cardiovascular risk factor medications after breast cancer diagnosis and incidence of cardiac events. Cancer. 2020;126:1541–9.
12. Hershman DL, Till C, Shen S, et al. Association of Cardiovascular Risk Factors With Cardiac Events and Survival Outcomes Among Patients With Breast Cancer Enrolled in SWOG Clinical Trials. J Clin Oncol. 2018;36:2710–7.
13. Lee DH, Fradley MG. Cardiovascular Complications of Multiple Myeloma Treatment: Evaluation, Management, and Prevention. Curr Treat Options Cardiovasc Med. 2018;20:19.
14. Lyon AR, Dent S, Stanway S, et al. Baseline cardiovascular risk assessment in cancer patients scheduled to receive cardiotoxic cancer therapies: a position statement and new risk assessment tools from the Cardio-Oncology Study Group of the Heart Failure Association of the European Society of Cardiology in collaboration with the International Cardio-Oncology Society. Eur J Heart Fail. 2020;22:1945-1960.
15. Neves KB, Rios FJ, van der Mey L, et al. VEGFR (Vascular Endothelial Growth Factor Receptor) Inhibition Induces Cardiovascular Damage via Redox-Sensitive Processes. Hypertension. 2018;71:638–47.
16. Piccirillo JF, Tierney RM, Costas I, et al. Prognostic importance of comorbidity in a hospital-based cancer registry. JAMA. 2004;291:2441–7.
17. Pinder MC, Duan Z, Goodwin JS, et al. Congestive Heart Failure in Older Women Treated With Adjuvant Anthracycline Chemotherapy for Breast Cancer. J Clin Oncol. 2007;25:3808–15.
18. Russo A, Novo G, Lancellotti P, Giordano A, Pinto FJ. Cardiovascular Complications in Cancer Therapy. Springer International Publishing Imprint, Humana Press; 2019. ISBN : 978-3-319-93401-3. https://doi.org/10.1007/978-3-319-93402-0
19. Sarocchi M, Arboscello E, Ghigliotti G, et al. Ivabradine in Cancer Treatment-Related Left Ventricular Dysfunction. Chemotherapy. 2018;63:315–20.
20. Spallarossa P, Maurea N, Cadeddu C, et al. A recommended practical approach to the management of anthracycline-based chemotherapy cardiotoxicity: an opinion paper of the working group on drug cardiotoxicity and cardioprotection, Italian Society of Cardiology. J Cardiovasc Med. 2016;17:e84.
21. Szmit S, Jurczak W, Zaucha JM, et al. Pre-existing arterial hypertension as a risk factor for early left ventricular systolic dysfunction following (R)-CHOP chemotherapy in patients with lymphoma. J Am Soc Hypertens JASH. 2014;8:791–9.

22. Tini G, Sarocchi M, Ameri P, et al. The Need for Cardiovascular Risk Factor Prevention in Cardio-Oncology. JACC Heart Fail. 2019;7:367–8.
23. Tini G, Sarocchi M, Sirello D, et al. Cardiovascular risk profile and events before and after treatment with anti-VEGF drugs in the setting of a structured cardio-oncologic program. Eur J Prev Cardiol. 2020:2047487320923056.
24. Tini G, Sarocchi M, Tocci G, et al. Arterial hypertension in cancer: The elephant in the room. Int J Cardiol. 2019;281:133–9.
25. Unger JM, Hershman DL, Fleury ME, et al. Association of Patient Comorbid Conditions With Cancer Clinical Trial Participation. JAMA Oncol. 2019;5:326–33.
26. Williams B, Mancia G, Spiering W, et al. 2018 ESC/ESH Guidelines for the management of arterial hypertensionThe Task Force for the management of arterial hypertension of the European Society of Cardiology (ESC) and the European Society of Hypertension (ESH). Eur Heart J. 2018;39:3021–104.
27. Yin AB, Brewster AM, Barac A, et al. Cardiovascular Prevention Strategies in Breast Cancer. JACC CardioOncology. 2019;1:322–5.

4

Cardiac Surgery and Percutaneous Coronary Intervention in Patients With Cancer

Fabrizio D'Ascenzo, MD, Phd, Stefano Salizzoni, MD, Ovidio De Filippo, MD, Guglielmo Gallone, MD, Francesco Bruno, MD, Mauro Rinaldi, MD., Gaetano Maria De Ferrari, MD.

Division of Cardiology, Department of Medical Science, Città della Salute e della Scienza, University of Turin, Italy; Division of Cardiac Surgery, Città della Salute e della Scienza, University of Turin, Italy

Correspondence to:
Fabrizio D'Ascenzo MD PhD, A.O.U. Città della Salute e della Scienza di Torino, Corso Bramante 88, 10126, Turin, Italy. Phone: +390116336023; FAX: +390116336769 Email address: fabrizio.dascenzo@unito.it

KEYWORDS: Acute Coronary Syndromes; Cancer; Percutaneous Coronary Interventions; Surgical Revascularization.

> *"It is not in the stars (in this case into perceived risk of cancer) to hold our Destiny but in ourselves (in this case into tailored therapy for cardiac problems)." (Julius Caesar, William Shakespeare)*

4.1 Introduction

4.1.1 The complex interplay between ACS and cancer: bench and pharmacological data

A complex interplay between ACS (acute coronary syndrome) and cancer exists due to common risk factors (like age, obesity, and smoking). Moreover, cancer itself and therapy may increase the risk of CAD, while the presence of coronary atherosclerosis may affect the management of these patients.

4.1.2 The complex interplay between ACS and cancer: the role of the disease

Cancer is widely described as a prothrombotic venous and arterial state and the incidence of arterial thrombosis is specifically higher in these patients (Al-Hawwas et al. 2018). Cancer cells increase the release of pro-inflammatory cytokines, which promote endothelial damage and enhances microvasculature permeability for pro-coagulating factors. Actually, arterial thrombosis in cancer can occur in the absence of an atherosclerotic plaque such as that observed in cardiovascular patients, where systemic hyper-coagulation is induced by several secreted factors from cancer cells, such as thrombin and vascular endothelial growth factor (VEGF), thereby promoting platelet activation and coagulation (Abdol Razak et al. 2018).

The classical concept of vascular thrombosis is that of Virchow's triad. Accordingly, a thrombus forms as a consequence of alterations of the blood contents (mainly platelets and coagulation factors), the vessel walls (mainly endothelium), and blood flow (mainly blood flow turbulence and stasis; see Figure 4.1). The first factor in Virchow's triad **(blood content)** appears critical to thrombogenesis in cancer patients. There is evidence that platelet

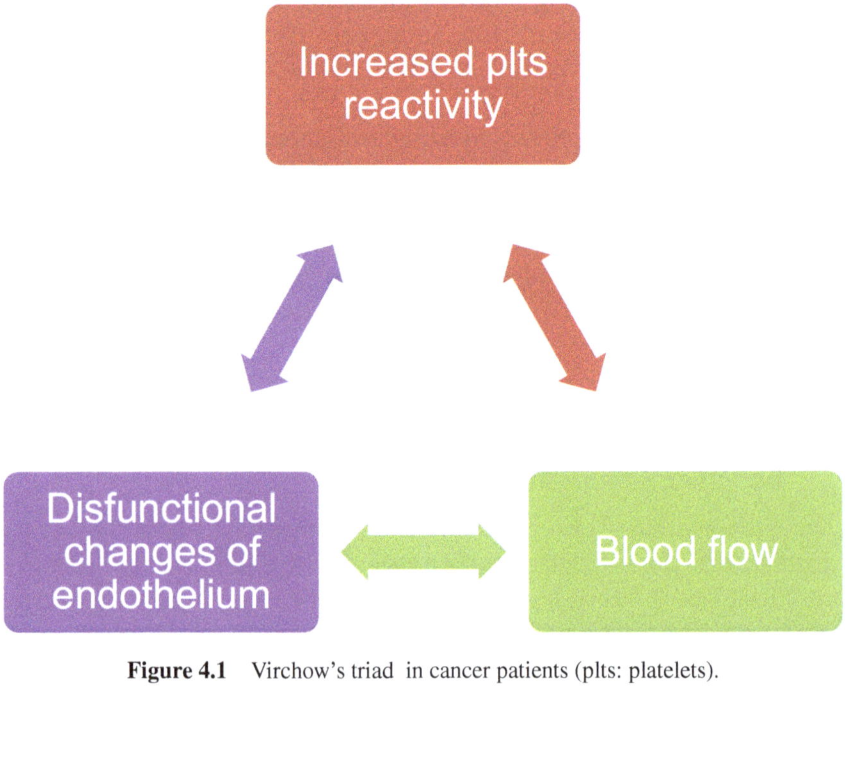

Figure 4.1 Virchow's triad in cancer patients (plts: platelets).

reactivity increase and that there are high circulating levels of platelet-specific products such as soluble P-selectin, platelet factor 4, thrombospondin, and beta-thromboglobulin (Sarkiss et al. 2007). Recently a bidirectional role has been demonstrated, with paraneoplastic cells activating platelets and, conversely, platelets enhancing cancer propagation and metastatic spread. Actually, in experimental models of pancreatic, colorectal, and renal tumors, a thrombocyte aggregation was noted after the interaction of platelets with tumor cells (Sarkiss et al. 2007).

The vascular wall, the second Virchow's triad element, also represents an important contributor to thrombosis in cancer patients. Indeed, cancer therapies, as we will discuss below, have a direct influence on the vascular wall, especially on the endothelium. Moreover, the interaction between platelets and subendothelium may be promoted by increased levels of von Willebrand factor (vWf; Sarkiss et al. 2007) triggered by inflammatory cytokines such as tumor necrosis factor and interleukin (IL)-1. Finally, loss of expression of thrombomodulin on the endothelial surface reduces the capacity to activate the anticoagulant protein C increasing prothrombotic state (Doll et al. 1986). These changes lead to endothelial dysfunction which fosters inflammation, proliferation, and vasoconstriction, all of which contribute to the development and clinical presentation of ischemic vascular disease (Sarkiss et al. 2007). The third factor in Virchow's trial, blood flow, appears to play a minor role in the etiology of thrombosis in cancer patients (Sarkiss et al. 2007).

4.1.3 The complex interplay between ACS and cancer: the role of chemotherapy and of radiotherapy

Many chemotherapeutic agents are known to be prothrombotic and there are multiple case reports documenting an association between chemotherapy and arterial thrombosis, although it is difficult to precisely define the relative prothrombotic effects of the chemotherapy compared to the hypercoagulable state of malignancy. Platinum-based agents (cisplatin), vascular endothelial growth factor (VEGF) inhibitors (bevacizumab), and VEGF tyrosine kinase receptor inhibitors (sorafenib/sunitinib/pazopanib) have been associated with increased rates of thrombosis. Moreover, endothelial dysfunction induced by bleomycin, endothelial apoptosis by vinblastine, and accelerated atherosclerosis by nilotinib and ponatinib have been demonstrated in pathology studies (Doll et al. 1986; FDA; Kosmas et al. 2008; Cortes et al. 2013). The most common arterial thrombotic events include myocardial

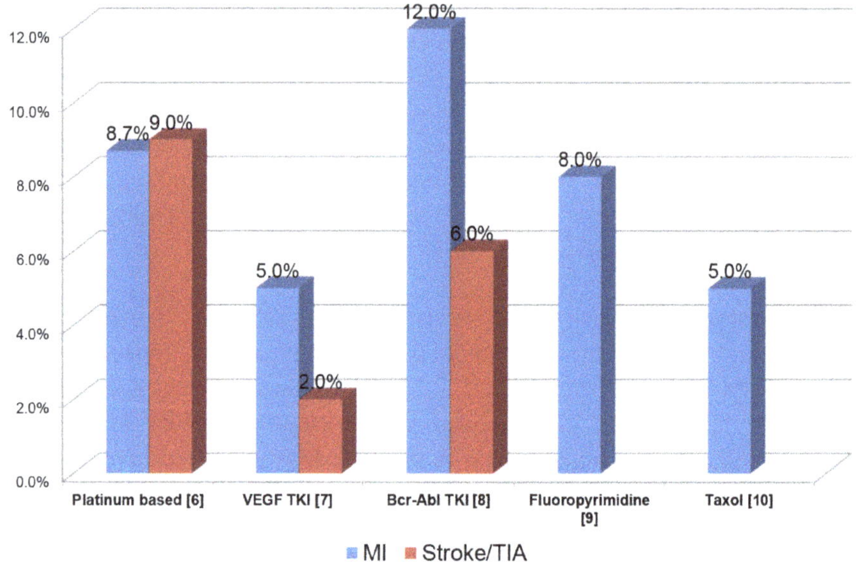

Figure 4.2 Incidence of MI and of stroke/TIA according to classes of drugs.

infarction (MI) and cerebrovascular events (stroke and transient ischemic attack); see Figure 4.2.

Interestingly, ionizing radiation affects not only cancerous but also non-cancerous cells, especially those that are rapidly proliferating, including endothelium (Hermann et al. 2016). Therefore, radiation therapy (RT), due to the oxidative stress and inflammation triggered, affects the coronary vasculature and may increase the risk of ACS. Even in the first month after initiation of RT, formation of cholesterol-rich plaques of abnormal intra-plaque hemorrhage and of thrombosis has been reported in animal models (Brosius et al. 1981). Clinically, these changes translate in fibrosis and later calcification in all three layers of the vessel wall (Brosius et al. 1981). In particular, the most exposed coronaries are the left anterior descending during left breast irradiation and the left main stem and the circumflex and the right coronary arteries during Hodgkin lymphoma treatment. Radiation-related cardiac disease in patients with lymphoma typically manifests 15–20 years after the initial treatment, and younger patients are more susceptible than older patients (van Nimwegen et al. 2016). For patients treated with RT for breast cancer, the risk of atherosclerotic plaques appears to increase linearly by increasing the whole heart radiation dose (Navi et al. 2017): for example, in a recent registry of women irradiated for breast cancer, the rate of major coronary events increased by 7.4%/Gy (Navi et al. 2017).

4.1.4 The complex interplay between ACS and cancer: clinical data

Patients with a malignancy pose several challenges when presenting with ACS. They are often older, with a large burden of comorbidities and present with a more relevant atherosclerotic burden. Their hematologic and coagulation abnormalities make the use of anticoagulants more difficult as that of antiplatelet agents and percutaneous coronary intervention (PCI). Moreover, patients with active malignancy, and often those with pre-existing cancers, have been excluded from randomized controlled trials or registries evaluating and defining the most innovative treatment for ACS. Furthermore, in all contemporary risk scores exploited to evaluate ischemic and bleeding risks, no data about cancer have been reported, although cancer diagnosis has far greater implications than the comorbidities included in these scores (Sarkiss et al. 2007; Marmagkiolis et al. 2016).

Regarding clinical presentation, cancer patients with ACS demonstrate an atypical profile compared to non-cancer patients. Less than one-third of cancer patients with ACS (30.3%) present with chest pain, 44% experience dyspnea, and 23% hypotension (Navi et al. 2017). The cause of this atypical presentation compared to non-cancer patients is undetermined. Strong analgesic therapy, neuropathy induced by the malignancy, chemotherapy, or radiation therapy could be the potential mechanisms. Therefore, it is important to have a higher clinical suspicion when screening cancer patients for ACS.

Different registries reported incidence of MI in patients with cancer. In 2017, Navi et al. (2017) analyzed the surveillance, epidemiology, and end results (SEER) database in order to evaluate the incidence of MI and stroke in cancer patients treated in the USA. A population of 279.719 patients with cancer was compared with the same number of patients without cancer. Interestingly in the first month after diagnosis, the incidence of MI was near 7% (being more than three times of that observed in non-cancer patients) declining over time (see Figure 4.3). Lung, colorectal, and pancreatic cancers were those more related with the risk of MI and increased also the risk of death. The same group investigated risk of MI for patients with occult cancer before diagnosis (Navi et al. 2019). They included more than 700.000 medicare beneficiaries, demonstrating that the risk starts increasing about five months before cancer diagnosis and thereafter progressively rise until peaking in the month before cancer diagnosis, when the risk was increased more than five-fold. Similar to the previous report, lung and colorectal cancers were those more related to an arterial thromboembolic event, especially for those patients

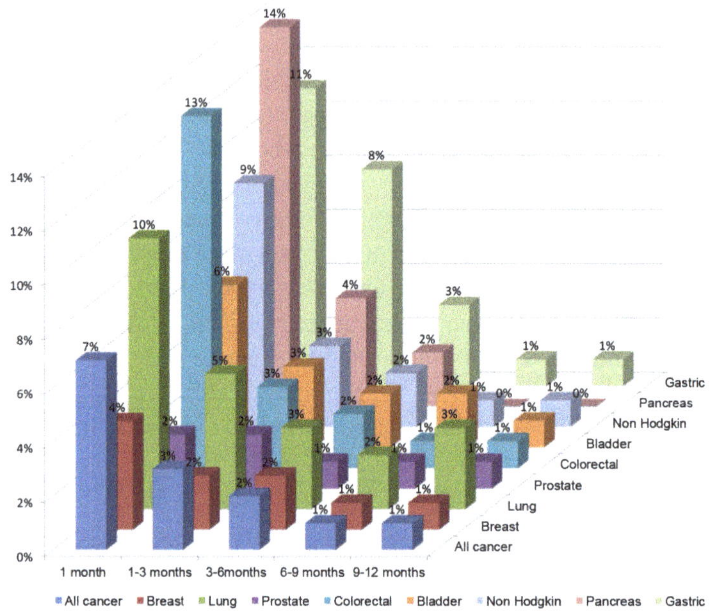

Figure 4.3 Incidence and timing of MI according to kind of cancer (Navi et al. [13]).

with stages III or IV cancer upon diagnosis. Interestingly, the demonstration that the risk of MI and ACS increases not only after diagnosis but also several months before stresses the hypothesis that the cancer's intrinsic effects on coagulation rather than as cancer treatment effects and antithrombotic interruption may be more relevant for etiology of arterial thrombosis. Finally, dual or single antiplatelet therapy interruption for diagnostic or interventional procedures, especially surgical biopsy of a newly discovered mass, might have contributed to the increased arterial thromboembolism risk before cancer diagnosis (D'Ascenzo et al. 2017).

Regarding current management of ACS patients with cancer, in a recent study including patients with ACS treated with clopidogrel, prasugrel, and ticagrelor (the BLEEMACS project), 926 patients with cancer and ACS (6.4%) were compared vs. 14.475 with ACS but non-cancer (Iannaccone et al. 2018). At the one year of follow-up, 3.7% of the patients had died, 11.7% in the cancer group and 3.2% in the non-cancer group ($P < 0.001$). Furthermore, patients with cancer more often experienced re-infarction (8.3% vs. 3.6%, $P < 0.001$) and bleedings (6.5% vs. 3%, $P < 0.001$) during follow-up. At the multiple regression analysis, the presence of cancer was the strongest independent predictor for the primary endpoint [hazard ratio

(HR) 2.1, 1.8–2.5, $P < 0.001$] and bleeding (HR 1.5, 1.1–2.1, $P = 0.01$). Moreover, in patients with cancer, a history of bleeding was more frequently observed, which explained the caution of physicians about the administration of antiplatelets, in particular new antiplatelet drugs and anticoagulant therapy. Ticagrelor and prasugrel were probably considered to be less safe than other antiplatelet agents such as aspirin or clopidogrel. Although a small sample, the new antiplatelet drugs seemed not to increase significantly the risk of bleeding and impact significantly on death. On the other hand, DAPT more than therapy with aspirin or clopidogrel alone seemed to have a protective effect. Similar results were reported in a large report of more than six million of patients (Bharadwaj et al. 2020): 186,604 with current cancer and more than 400,000 patients with a previous diagnosis. Interestingly, patients with previous cancers and especially those with current cancer were treated less frequently with an invasive approach and were exposed to a risk of death two times higher and a risk of bleeding three times higher than non-cancer patients.

These last two papers perfectly resume the most relevant problems of cancer patients presenting with ACS:

1. the perceived risk of “futile” procedures consequently limiting potentially life-saving interventions;
2. the high risk of bleeding which, regarding PCI (percutaneous coronary interventions), in the last year has been partially solved by introduction of coronary stents requiring short-term DAPT (D’Ascenzo et al. 2017);
3. the need of invasive diagnostic or therapeutic non-cardiac procedures requiring DAPT interruption which, despite innovative protocols based on cangrelor, remains validated on small sample size of patients (Rossini et al. 2018).

4.1.5 The complex interplay between cardiac surgery and cancer: clinical data

The evaluation of patients with a surgical indication is deeply influenced by reports showing a close relationship between the intervention and the progression of the disease (Mistiaen et al. 2004). Use of a cardiopulmonary bypass (Markewitz et al. 1996; Payen et al. 2000) was demonstrated to induce an inflammatory reaction, including an increased release of cytokines and a decrease in activity of interleukin-2 and interleukin-12, which are of importance in the regulation of cell-mediated immunity. After cardiac surgery, a shift of T-cell phenotype toward humoral immunity has been observed, with a concomitant suppression of cell mediated immunity. T cells with a CD4 expression were also decreased, up to seven days after the operation. Some of these cytokines had also been involved in the immunologic defense against tumor cell formation, invasion, and metastasis. Actually, the paper of Miasten et al. (2004) demonstrated that a fatal progression of the tumor is seen if the time interval between the occurrence of the malignant tumor and cardiac surgery is shorter than two years, potentially confirming the previous hypothesis. Recently, in a report of the SWEDEHEART registry (Mistiaen et al. 2004) on the long-term mortality of 82,137 patients undergoing CABG between 1997 and 2015, of which 6819 procedures (8.3%) were performed in patients with a history of cancer, the proportion increased over time from 3.8% in 1997 to 14.8% in 2015. The most frequent cancers involved were: prostate (31%), the gastrointestinal ones (12%), those of the kidneys (11%), and breast (8%). An increased and adjusted long-term mortality risk was reported in patients with a history of cancer that was inversely associated with the time interval from cancer diagnosis to CABG. The authors calculate cutoff point of 2.5 years for the time period between cancer diagnosis and CABG beyond which history of cancer was not significantly associated with all-cause mortality.

Another high-risk subset of patients is represented by those with severe aortic stenosis (AS). Tailoring the correct and most optimal care for these patients is complex. Actually, aortic valve replacement or transcatheter aortic valve implantations (TAVI) are the only ways to reduce mortality in these patients (D'Ascenzo et al. 2013a, 2013b; Conrotto et al. 2015; D'Onofrio et al. 2020). On the other hand, the ongoing improvement of cancer treatment may lead these patients to be more threatened by the AS (if left untreated) than by the cancer (Swain et al. 2013). Moreover, they may be limited by AS due to the augmented risk of oncologic surgery and risks associated with cardio toxicity and heart failure (Minamino-Muta et al. 2018).

However, for patients with ACS, severe coronary artery disease amenable to surgical revascularization, or severe AS, the determination of individual patient life expectancy and ascertaining benefit vs. futility remains crucial, not to deny to these patients life-saving intervention.

In conclusion, as stated by Julius Caesar in the tragedy of William Shakespeare, it is not in the stars (in this case, into perceived risk of cancer) to hold our Destiny but in ourselves (in this case, into tailored therapy for cardiac problems) (Shakespeare 1599).

References

1. Abdol Razak NB, Jones G, Bhandari M, et al. Cancer-Associated Thrombosis: An Overview of Mechanisms, Risk Factors, and Treatment. Cancers (Basel). 2018;10:380.

2. Al-Hawwas M, Tsitlakidou D, Gupta N, et al. Acute Coronary Syndrome Management in Cancer Patient Curr Oncol Rep. 2018;20:78.
3. Bharadwaj A, Potts J, Mohamed MO, et al. Acute myocardial infarction treatments and outcomes in 6.5 million patients with a current or historical diagnosis of cancer in the USA. Eur Heart J. 2020;41:2183-2193.
4. Brosius FC 3rd, Waller BF, Roberts WC. Radiation heart disease. Analysis of 16 young (aged 15 to 33 years) necropsy patients who received over 3,500 rads to the heart. Am J Med. 1981;70:519-30.
5. Conrotto F, D'Ascenzo F, Presbitero P, et al. Effect of gender after transcatheter aortic valve implantation: a meta-analysis. Ann Thorac Surg. 2015;99:809-16.
6. D'Ascenzo F, Ballocca F, Moretti C, et al. Inaccuracy of available surgical risk scores to predict outcomes after transcatheter aortic valve replacement. J Cardiovasc Med (Hagerstown). 2013a;14:894-8.
7. D'Ascenzo F, Chieffo A, Cerrato E, et al. Incidence and Management of Restenosis After Treatment of Unprotected Left Main Disease With Second-Generation Drug-Eluting Stents (from Failure in Left Main Study With 2nd Generation Stents-Cardiogroup III Study). Am J Cardiol. 2017;119:978-982.
8. D'Ascenzo F, Gonella A, Moretti C, et al. Gender differences in patients undergoing TAVI: a multicentre study. EuroIntervention. 2013b;9:367-72.
9. D'Ascenzo F, Iannaccone M, Saint-Hilary G, et al. Impact of design of coronary stents and length of dual antiplatelet therapies on ischaemic and bleeding events: a network meta-analysis of 64 randomized controlled trials and 102 735 patients. Eur Heart J. 2017;38:3160-3172.
10. D'Onofrio A, Salizzoni S, Filippini C, et al. Surgical aortic valve replacement with new-generation bioprostheses: Sutureless versus rapid-deployment J Thorac Cardiovasc Surg. 2020;159:432-442.e1
11. Herrmann J, Yang EH, Iliescu CA, et al. Vascular toxicities of cancer therapies: the old and the new–an evolving avenue. Circulation. 2016;133:1272–89
12. Iannaccone M, D'Ascenzo F, Vadalà P, et al. Prevalence and outcome of patients with cancer and acute coronary syndrome undergoing percutaneous coronary intervention: a BleeMACS substudy Eur Heart J Acute Cardiovasc Care. 2018;7:631-638.
13. Markewitz A, Faist E, Lang S, Hultner L, Weinhold C, Reichart B. An imbalance in T-helper cells subsets alters immune response after cardiac surgery. Eur J Cardiothorac
14. Minamino-Muta E, Kato T, Morimoto T, et al.; CURRENT AS registry Investigators. Malignant disease as a comorbidity in patients with severe

aortic stenosis: clinical presentation, outcomes, and management. Eur Heart J Qual Care Clin Outcomes. 2018;4:180-188.

15. Mistiaen WP, Van Cauwelaert P, Muylaert P, et al. Effect of prior malignancy on survival after cardiac surgery. Ann Thorac Surg. 2004;77:1593-7
16. Mistiaen WP, Van Cauwelaert P, Muylaert P, et al. Effect of prior malignancy on survival after cardiac surgery. Ann Thorac Surg. 2004; 77:1593-7.
17. Navi BB, Reiner AS, Kamel H, et al. Arterial thromboembolic events preceding the diagnosis of cancer in older persons. Blood. 2019;133: 781-789.
18. Navi BB, Reiner AS, Kamel H, et al. Risk of arterial thromboembolism in patients with cancer. J Am Coll Cardiol. 2017;70:926–38.
19. Payen D, Faivre V, Lukaszewicz AC, et al. Assessment of immunological status in the critically ill. Minerva Anestesiol 2000;66:757–63.
20. Rossini R, Tarantini G, Musumeci G, et al.; Italian Society of Interventional Cardiology (SICI-GISE); Italian Society for the Study of Haemostasis and Thrombosis (SISET); Italian Society of Anesthesia and Intensive Care Medicine (SIAARTI); Italian Society of Surgery (SIC); Italian Society for Cardiac Surgery (SICCH); Italian Society of Vascular and Endovascular Surgery (SICVE); Italian Society of Urology (SIU); Italian Orthopaedic Society (SIOT); Italian Society of Thoracic Surgeons (SICT); Italian Federation of Scientific Societies of Digestive System Diseases (FISMAD); Italian Society of Digestive Endoscopy (SIED); Italian Association of Hospital Gastroenterology and Digestive Endoscopy (AIGO); Italian Association of Gastroenterology and Digestive Endoscopy (SIGE); Italian Society of Maxillofacial Surgery (SICMF); Italian Society of Reconstructive Plastic Surgery and Aesthetics (SICPRE); Italian Society of Gynecology and Obstetrics (SIGO); Italian Society of Neurosurgery (SINch); Italian Association of Hospital Pulmonologist (AIPO); Italian Society of Periodontology (SIdP); Italian Society of Ophthalmology (SOI); Italian Association of Hospital Otorhinolaryngologist (AOOI); Italian Association of Hospital Surgeons (ACOI); Association of Obstetricians Gynecologists Italian Hospital (AOGOI) A Multidisciplinary Approach on the Perioperative Antithrombotic Management of Patients With Coronary Stents Undergoing Surgery: Surgery After Stenting JACC Cardiovasc Interv. 2018;11:417-434
21. Shah K, Gupta S, Ghosh J, et al. Acute non-ST elevation myocardial infarction following paclitaxel administration for ovarian carcinoma: a case report and review of literature. J Cancer Res Ther. 2012;8:442–4

22. Shakespeare W. 1599. https://william-shakespeare.classic-literature.co.uk/the-tragedie-of-julius-caesar/
23. Swain SM, Kim SB, Corte ́s J, et al. Pertuzumab, trastuzumab, and docetaxel for HER2-positive metastatic breast cancer (CLEOPATRA study): overall survival results from a randomised, doubleblind, placebo-controlled, phase 3 study. Lancet Oncol 2013;14:461–471
24. van Nimwegen FA, Schaapveld M, Cutter DJ, et al. Radiation Dose-Response Relationship for Risk of Coronary Heart Disease in Survivors of Hodgkin Lymphoma. J Clin Oncol. 2016;34:235-43.

5

Cancer Therapy-Induced Arrhythmias

Davide Castagno MD, PhD*; Vincenzo Cusenza MD*, Gaetano Maria De Ferrari, MD

*Division of Cardiology, Department of Medical Sciences, "Città della Salute e della Scienza Hospital," University of Turin, Italy.

Correspondence to:
Prof Davide Castagno, MD PhD
Division of Cardiology, Department of Medical Sciences,
Città della Salute e della Scienza Hospital,
University of Turin, Turin, Italy.
Email: davide.castagno@unito.it

KEYWORDS: Arrhythmias; Cardiomyopathy; Heart Failure; Chemotherapy; Antineoplastic Agents; Neoplasm.

5.1 Introduction

5.1.1 Epidemiology

Cancer therapies are associated with a broad range of cardiovascular toxicities. Although the initial focus in cardio-oncology has been on early detection and prevention of heart failure, there is increasing awareness of the arrhythmogenic potential of cancer therapies. Both the improvement in prognosis and the development of new targeted anticancer therapies have contributed to the increased prevalence of cardiac arrhythmias in patients with cancer (Zamorano et al. 2016). However, it is likely that the incidence of cardiac arrhythmias has been largely underestimated in the past because of poor (or absent) screening strategies and due to exclusion from key oncological trials of patients with pre-existing heart disease (Chang et al. 2017). Supraventricular tachycardias, particularly atrial fibrillation, are frequent in patients undergoing cancer treatment. The incidence of atrial fibrillation among patients receiving active cancer therapy ranges from 2% to 16% depending on the cohort studied and it is associated with two-fold increased risk of thromboembolic events and six-fold increased risk of heart failure (Farmakis et al. 2014; Hu et al. 2013).

Another frequent and potential pro-arrhythmic complication in patients with cancer is QT interval prolongation. Multiple factors, including direct effect of chemotherapy on potassium channels and/or intracellular signal pathways, electrolyte derangements, and use of antiemetics and antibiotics can facilitate QT interval prolongation. In a study evaluating ECG abnormalities in patients enrolled in a phase I clinical trial, QT interval prolongation occurred in 20% of cases, but arrhythmic events were clinically insignificant (Naing et al. 2012). Arrhythmic complications and the occurrence of life-threatening arrhythmias are usually limited to significant QT interval changes (more than 60 ms prolongation versus baseline QT) or very long QT duration (more than 500 ms) (Brell et al. 2010).

Cardiac conduction disturbances and brady-arrhythmias are less common and limited to specific antineoplastic agents. Asymptomatic bradycardia can occur in up to 30% of patients receiving paclitaxel, whereas heart block has been observed in 0.1% of cases (Arbuck et al. 1993). Another drug significantly associated with bradycardia, especially in combination with other drugs with negative chronotropic effect (i.e., beta-blockers, calcium-channel blockers, and digoxin), is thalidomide (Tamargo et al. 2015). Also targeted cancer therapies, such as multi-target kinase inhibitors and histone deacetylase inhibitors, have been associated with sinus bradycardia in up to 15% of patients with rarer cases of sinus arrest and asystole (Herrmann 2020). In patients receiving immune check point inhibitor therapies, 10% of cardiotoxic manifestations are atrio-ventricular blocks and conduction disturbances (Mir et al. 2018).

5.1.2 Etiology and Pathogenesis

Three main mechanisms can be recognized in the pathogenesis of cancer-related arrythmias:

- direct effect on ionic channels;
- indirect effects on the heart mediated by systemic alterations;
- myocyte and/or conduction system cells damage.

1. Direct effect on ionic channels

   The most important secondary effect of the interaction between chemotherapies and ionic channels is QTc prolongation with potential subsequent major arrhythmias (i.e., torsade de pointes, ventricular tachycardia, ventricular fibrillation, and sudden death). Among ionic channels, the most frequently affected is IKr (potassium rapid inward rectifier, commonly known as hERG). Other ionic channels frequently affected are IKs (potassium slow inward rectifier), ATP-dependent K^+ current (e.g., by arsenic trioxide), Na^+ inward current (e.g., by oxaliplatinum), Na^+/Ca^{++} exchanger (e.g., by adriamycin) (Guglin et al. 2009).

2. Indirect effects on the heart mediated by systemic alterations
Electrolyte imbalances can facilitate and beget cardiac arrhythmias by altering membrane potential. Such alterations are commonly encountered in cancer patients. Sustained diarrhea, vomiting, and sweating often lead to ions loss (e.g., sodium, potassium, chloride, and magnesium); tumor lysis syndrome can lead to hyperkalemia, hyperphosphatemia, and hypocalcaemia; kidney damage, chemotherapies-related hyporexia, drug stimulated hemopoiesis, and various types of paraneoplastic syndromes are all causative of ionic derangements (Berardi et al. 2019). Moreover, drugs used to mitigate cancer-therapy side effects can be arrhythmogenic too (e.g., some anti-emetics prolong QTc) (Zamorano et al. 2016).
Among chemotherapy-induced systemic disorders, also endocrine disturbances can be associated with arrhythmias. Hypothyroidism, a possible side effect of thalidomide (frequently used in multiple myeloma therapy) or sunitinib (used for treatment of renal cell carcinoma or for treatment of gastro intestinal stromal tumor) and other drugs as imatinib, axitinib, or interferon-alpha, can cause significant sinus bradycardia. Nevertheless, reduced thyroid function has been also linked with a decreased probability of cardiac arrythmias secondary to the increase of the arrhythmogenic threshold (Osuna et al. 2017). Hypoparathyroidism is a very rare side effect of chemotherapies (e.g., vinca alkaloid) that can lead to hypocalcaemia (Ajero et al. 2010). Lastly, adrenal insufficiency can cause complex ions disorders (hyperkalemia, hypercalcemia, and hyponatremia) and should be considered mostly in patients treated with a long course of steroid. All the above discussed systemic disorders could also either be part of the clinical presentation of a paraneoplastic syndrome or the direct consequence of tumor growth.

3. Myocyte and/or conduction system cells damage
This is probably the most frequent cause of chemotherapy related arrhythmias. Myocytes direct damage can cause various degrees of cardiomyopathy with or without heart failure symptoms. Even in the context of a completely normal functioning conduction system, this can lead to arrhythmias by slowing down of electrical signal and mechanism

of re-entry after the development of scar and fibrous zones. In addition, chemotherapy-induced ischemia can contribute to development of arrhythmias similarly to what is commonly observed in a patient suffering from coronary artery disease.

The huge variability in chemotherapy mechanisms of action reflects in three different types of damage (Herrmann 2020):

- type 1 (primary): it is the prototype of direct cardiotoxicity. Chemotherapeutics causing this kind of damage are usually called "conventional" and act by killing neoplastic cells taking advantage of their high metabolic demand and activity and/or inducing oxidative stress. Unfortunately, this process is not selective on tumor cells, leading to well-known side effects. Cardiac toxicity caused by anthracyclines, which have shown to selectively act on cardiomyocyte mitochondria in experimental models, is the most frequent form of type 1 damage encountered in clinical practice. However, anthracyclines related arrhythmias are relatively rare (about 3%). In contrast, a strong association between taxanes and clinically significant bradycardia has been previously observed (up to 30% in patients receiving paclitaxel) with rare cases of atrioventricular block (of all degrees). The mechanisms underlying these phenomena have been poorly elucidated so far (Herrmann 2020). These drugs are obtained from yew tree (*taxus baccata*), rich in taxine alkaloids that act blocking Na^{+}/Ca^{++} channels (Natasha et al. 2019). Irritative direct effects on heart structure are also included in this subtype of chemotherapy mediated cardiac damage. With this regard, neoplastic infiltration and direct mechanical pressure overload can manifest clinically as atrial fibrillation which is a common side effect of chemotherapeutics like cisplatin, especially when administered intra-pericardially (Tomkowski et al. 1997; Richards et al. 2006; Dhesi et al. 2013).
- type 2 (secondary or indirect): this type of damage is frequently mediated by alterations in perfusion and, probably less commonly, by changes in endocrine (already discussed above) and innervation milieu. 5-Fluoro-uracile and its prodrug capecitabine are both known to produce profound and diffuse vasoconstriction and vasospasms leading to myocardial infarction or Takotsubo syndrome.
- VEGF inhibitors (such as Bevacizumab and Ranizumab), by preventing angiogenesis, can worsen any pre-existing vascular disease and exacerbate the effect of well-known cardiovascular risk factors such as hypertension. This side effect has been observed also in some

tyrosine kinase inhibitors (e.g., sorafenib and sunitinib) although their pleiotropic biological activity makes unlikely that VEGF inhibition is the only mechanism involved. Ischemia can also be the consequence of direct infiltration of coronary arteries by adjacent neoplasm. Of note, ischemia can cause QTc prolongation, facilitating the occurrence of major arrhythmias. Finally, the compression of neural structures by solid tumors (e.g., compression of vagus nerve by neck tumor) can exert significant cardio-inhibitory effects making pacemaker implantation mandatory in cases of impossible removal/decompression.

- type 3 (inflammatory cell infiltration): chemotherapeutics can induce myocardial inflammation and myocarditis. Ensuing myocytes damage and disruption of specialized conduction system can lead to ventricular tachycardias, heart blocks, and sudden death. A classic example is cyclophosphamide induced hemorrhagic myocarditis, characterized by capillary micro-thrombosis with fibrin deposition (Dhesi et al. 2013). Anyway, it should be taken into account that there is significant overlap between classifications; for example, cyclophosphamide can also determine type 1 cardiotoxicity, while anthracyclines carry significant risk of type 3 cardiotoxicity (besides type 1 cardiotoxicity). Recently, the use of immune checkpoint inhibitors (ICI) has been associated with clinically relevant myocarditis, presenting in severe rare cases as cardiogenic shock and sudden death. The proposed pathophysiological mechanisms are generation of autoantibodies, T cell (re)activation, and production of inflammatory cytokines (Sury et al. 2018).

5.1.3 Diagnosis, Physical Examination, and Diagnostic Tests

As mentioned before, arrhythmias are probably underdiagnosed in cardio-oncology since continuous monitoring is rarely performed or limited to non-continuous 12 leads ECG9. Furthermore, patients with pre-existing cardiovascular disease and possible substrate for cardiac arrythmias have been consistently excluded from major oncological trials (Zamorano et al. 2016; Lopez Fernandez and Van der Meer 2019). However, there is reason to believe that in the near future, mobile health technologies will dramatically improve early detection of arrhythmias, in particular of atrial fibrillation, possibly improving morbidity and mortality of oncological patients (Hindricks et al. 2020). That being said, there are no particular hallmarks in the diagnosis of cancer-related arrhythmias. The best way to diagnose arrhythmias is indeed to identify patients at risk and to intensify monitoring strategies; as a general rule, common cardiovascular risk factors should be identified (and if possible treated) throughout all the clinical journeys of oncological patients. Anamnestic collection focused on history of arrhythmias, structural heart disease, or predisposing substrate is of paramount importance. Timing of follow-up should be tailored according to patient's risk profile, pre-existing clinical conditions, and the arrhythmogenicity of chemotherapy used. During general examination, irregular heartbeat should be carefully investigated either by checking pulse or auscultating the heart. In rare occasions, the diagnosis of arrhythmia is made in the emergency setting, especially in the case of hemodynamic compromise or cardiac arrest requiring urgent medical intervention. Twelve-lead ECG, ECG Holter monitoring, and implantable loop recorder are the most frequently used diagnostic tools. Some arrhythmias can manifest more frequently under stress; therefore, the use of an exercise test may prove extremely useful. Electrophysiological study is the diagnostic gold standard but is rarely performed, considering its invasiveness. The choice of the right diagnostic tool should be taken according to symptoms frequency, suspected arrhythmia, and patient's characteristics. There are no significant differences in the use of diagnostic tests in oncological patients, but arrhythmias can be temporally correlated with chemotherapy cycle; thus, performance and timing of investigation should be set accordingly (Lopez Fernandez and Van der Meer 2019).

Twelve-leads ECG is mandatory to assess basal condition (e.g., heart rate, PR, QRS, QTc interval duration, etc.) and to ascertain the presence of abnormalities (e.g., significant bradycardia or tachycardia, conduction blocks, Q waves, repolarization disorders, etc.) in patients treated (or planned for treatment) with chemotherapeutics with known arrhythmogenic side effects. Since QTc interval prolongation is probably the most commonly encountered ECG modification in the oncological setting, it is worthy to specify how its measure and calculation should be performed. While Bazett formula is the most widely used for QT interval correction, its use in cardio-oncology has been questioned. Fridericia and Framingham formulas (but also Hodges and Rautaharju) have shown better rate correction compared with Bazett formula which has significant risk of overcorrection at high heart rates and undercorrection at lower heart rates (Vanderberk et al. 2016; Muluneh et al. 2019). Good clinical practice is to use the same formula used in original chemotherapeutics clinical trials. Automatic machine calculations often provide longer QTc by measuring from the earliest QRS onset of all leads to the latest offset of all leads. Upper normal limits for QTc are 450 ms in men and 460 ms in women; however, 500 ms and $\Delta QT > 60$ ms are the cutoff of concern since, beyond these values, the risk of torsade de point increases significantly (Priori et al. 2015). Finally, conduction abnormalities such as complete bundle branch block require adjustment of the measured QT (e.g. subtracting the exceeding QRS duration due to intraventricular conduction delay from measured QT). This substraction is done before applying the formula chosen for correction or, alternatively, a different cutoff level (e.g., 550 ms) can be used (Porta-Sánchez et al. 2017; Bogossian et al. 2020).

Regardless of the suspected etiology of the arrhythmia, electrolytes, full blood count, and thyroid function should always be tested (Zamorano et al. 2016; Lopez Fernandez and Van der Meer 2019). Echocardiogram is warranted to assess eventual structural disease (Zamorano et al. 2016; Brugada et al. 2019; Lopez Fernandez and Van der Meer 2019; Hindricks et al. 2020). Further specific investigations are necessary in case of arrhythmias encountered in peculiar clinical scenarios. For example, when myocarditis is suspected, cardiac MRI, and, in second instance, endomyocardial biopsy, can improve diagnostic power (Zamorano et al. 2016). Coronary-CT, SPECT, functional MRI, and invasive coronary angiography allow to detect CAD. With this regard, it should always be remembered that radiotherapy exposure, especially at younger age and high doses, is associated with proximal coronary involvement increasing the risk of large myocardial infarctions. Therefore, a lower threshold for invasive investigations in patients deemed at risk is reasonable (Hu et al. 2013).

5.1.4 Clinical Characteristics

Several kinds of arrhythmias can be observed in patients with cancer. As in the general population, atrial fibrillation is the most common cause of concern. While the pro-arrhythmic effect of older chemotherapeutics (e.g., anthracyclines) is well known, the widespread use of newer compounds (e.g., TKI, immunomodulators, etc.) has increased the clinical scenario complexity. For example, evaluation for atrial fibrillation occurrence in patient treated with ibrutinib requires close collaboration between hematologist and cardiologist. Moreover, it should be considered that some arrhythmias are associated with combination of different drugs rather than with a single agent and that often they are facilitated by electrolyte derangements (sometimes caused by the same drug or concomitant agent used). In the subsequent table, we provide a list of chemotherapy drugs used in clinical practice and the arrhythmias they are more frequently associated with.

5.1.5 Treatment

Treatment of chemotherapy side effects, including arrhythmic complications, often represents a hard clinical challenge. As in other field of oncology,

Therapy Class	Agent	AF	SVT	Bradycardia	AV Block	QTc prolongation	TdP	VT/VF	SCD
Inorganic compound	Arsenic trioxide	++	++	+	+	+++	++	+	+
Alkylating agent	Anthracyclines[a]	+++	++	+		++[b]		+	+
	Amsacrine		+			+		+	+
	Cyclophosphamide	+		cr[c]	cr			cr	
	Ifosfamide	+	+	+				+	cr
	Melphalan	+++	++					cr	
Antimetabolites	5-FU	+	cr	+++	cr	cr	cr	++	++
	Capecitabine	++	cr	++	cr	+++	cr	+	+
	Cytarabine[c]	cr							
	Fludarabine		++	cr					
	Pentostatin[c]			cr					cr
	Gemcitabine	+	+					cr	
	Clofarabine	+++[d]	cr	cr		cr			
Antifolate	Methotrexate		+/-					+/-	
Microtubule agents	Paclitaxel[e]	+	+	+++	+			+	cr
	Docetaxel[c]	cr	cr		+				cr
	Vincristine[c]	cr							
Platinum-based drugs	Cisplatin[f]	+	+	+	+			+	
Immunomodulatory drugs	Lenalidomide	++		cr				cr	cr
	Thalidomide	+	+	+++	+			cr	cr
Endogenous molecule	Interleukin-2	+	+	+	cr			+	cr
	Interferon-alfa		+					+	+/-
Proteasome inhibitors	Bortezomib	cr	cr	+	cr	cr	cr	cr	cr
	Carfilzomib[g]	cr		cr	cr				+
HDAC inhibitors	Romidepsin	+	++			++	+	++	+
	Panobinostat	cr	cr	cr		++	cr	cr	cr
	Vorinostat	cr		cr		++	cr		cr
CDK4/6 inhibitors	Ribociclib					++			
mTOR inhibitors	Everolimus	++							

Therapy Class	Agent	AF	SVT	Bradycardia	AV Block	QTc prolongation	TdP	VT/VF	SCD
Monoclonal antibodies	Alemtuzumab	++		++				+	+
	Cetuximab	+		+				+	+
	Necitumumab		+						++
	Pertuzumab	+	+	+				+	+
	Rituximab	+	+	+	+	+	+	+	+ [h]
	Trastuzumab	++	++	+				+	cr [h]
Kinase inhibitors	Osimertinib					++			
	Lapatinib	+	+			+			cr
	Lenvatinib					++			
	Pazopanib			+++		++			
	Sorafenib	+		+	+	+	+		
	Sunitinib	cr		+		+	+		cr
	Vandetanib					+++		+	+
	Bosutinib			+		++			
	Dasatinib	+	+			+		+	+
	Imatinib	+	+						
	Nilotinib	++		++	++	++			+
	Ponatinib	++	+	+	+	+		+	
	Ibrutinib[i]	+++	+					+	+
	Zanubrutinib	+							
	Alectinib			+++		+			
	Ceritinib			+		++			
	Crizotinib			+++		+			
	Brigatinib			++					
	Lorlatinib				+				
	Encorafenib					+			
	Vemurafenib	++	+	+		+++	+		
	Gilteritinib					++			
	Trametinib			++		++			
	Ruxolitinib			+		+			
	Selpercatinib					+			

Therapy Class	Agent	AF	SVT	Bradycardia	AV Block	QTc prolongation	TdP	VT/VF	SCD
Immune checkpoint inhibitors	Ipilimumab	+		+	+			+	+
	Nivolumab	+		+	+			+	+
	Pembrolizumab	+		+	+			+	+
CAR-T cell therapy	Tisagenlecleucel	++	++						

a) Arrhythmic toxicity, contrary to cardiomyopathy, seems to be dose-independent (Buza et al. 2017).
b) QT dispersion can be reduced by dexrazoxane (Galetta et al. 2005).
c) Evidence limited to case report.
d) If combined with cytarabine.
e) Associated also with right and left bundle branch block (Kamineni et al. 2003).
f) Risk of AF arises significantly if used for intrapericardial instillation or hyperthermic abdominal lavage (Tomkowski et al. 1997; Richards et al. 2006; Dhesi et al. 2013).
g) The most part of cardiac adverse effect occurred during the first cycles and frequency did not increase after (Siegel et al. 2013).
h) SCD could possibly be related to severe hypomagnesemia (Buza et al. 2017).
i) Ibrutinib is associated with enhanced risk of major bleeding too, and appropriate anticoagulation therapy can be challenging (Lipsky et al. 2015).

the inherent difficulty is to balance prevention of unacceptable side effects avoiding interruptions of potential life-saving therapies. Close collaboration between cardiologists and oncologists is of paramount importance to deliver the best of care.

5.1.5.1 Atrial fibrillation

Oversimplifying, atrial fibrillation treatment can be reduced to two main cornerstones: symptoms improvement and stroke prevention.

- Symptoms improvement
  There is no compelling evidence of a significant difference in terms of mortality between a rhythm control and rate control strategy in patients with atrial fibrillation (Hindricks et al. 2020). The recent EAST – AFNET 4 trial showed that early initiation of rhythm-control therapy may convey clinical benefit in patients with recently diagnosed atrial fibrillation (Kirchhof et al. 2020). Active cancer or history of neoplasm were not among the exclusion criteria of the trial, but patients with any disease limiting life expectancy to less than 1 year could not be enrolled. Therefore, it is not clear whether the results of EAST – AFNET 4 trial can be extended to all patients with cancer developing atrial fibrillation and further investigations are warranted.
  The choice between rate and rhythm control should be mainly based on the aim of improving quality of life. Although this choice should follow usual considerations even in patients with cancer, two specific concerns should be taken into account. First, rhythm control can be more difficult to achieve since cancer itself, predisposing to development of atrial fibrillation, represents a non-removable potent arrhythmogenic factor. Thus, rate control is generally easier to achieve and favored over rhythm control (Lopez Fernandez and Van der Meer 2019; Hindricks et al. 2020). On the other hand, in selected patient (e.g., extremely symptomatic or with hemodynamic compromise) may be advisable to pursue rhythm control preventing atrial fibrillation from maintaining itself. Together with clinical judgment, patient's involvement play a central role in the decisional process. It is worth to remind that obtaining information about potential underlying structural and/or valvular heart disease is of paramount importance, especially in oncological patients, before initiating antiarrhythmic drugs or performing invasive procedures. Whenever rate control is the strategy of choice, it seems reasonable to maintain heart rate below 110 bpm at rest although no specific target has been defined yet. The RACE II study compared lenient rate control (resting heart rate < 110 bpm) vs. strict rate control (target resting heart

rate <80 bpm and <110 bpm during moderate exercise) in patients with permanent atrial fibrillation without showing significant differences regarding the development of cardiovascular morbidity, mortality, symptoms, quality of life, and atrial and ventricular remodeling (Van Gelder et al. 2010). In selecting the drug to achieve rate control, it should be remembered that both diltiazem and verapamil can inhibit CYP3A4, thus requiring close monitoring of drug–drug interactions. It is also important to remind that rapid atrial fibrillation may not always be the result of poor rate control. Oncological patients are indeed unfortunately prone to medical conditions that increase heart rate such as pain, hyperemesis, and dehydration. The same holds true for other symptoms such as fatigue and asthenia; the correct attribution of these symptoms may be challenging and an attempt of cardioversion and sinus rhythm restoration to assess their reversibility can represent a valid approach (Lopez Fernandez and Van der Meer 2019). Finally, to the best of our knowledge, the feasibility and efficacy of transcatheter ablation of atrial fibrillation in patients with active cancer has been poorly investigated so far. Surely this therapeutic approach has been increasingly used during the last two decades and should be kept in mind as a possible option in the field of cardio-oncology (Lopez Fernandez and Van der Meer 2019). However, the frailty and the susceptibility to bleeding of patients with cancer may represent a limit to its widespread use (Lopez Fernandez and Van der Meer 2019; Giustozzi et al. 2020).

- Thromboembolic prevention

  Patients with cancer have an increased risk of thrombotic (four- to seven-fold) and bleeding (two-fold) events (Prandoni et al. 2002; Fuentes et al. 2016). This notwithstanding, neither CHA2DS2VASc nor HAS-BLED score, which are commonly used to estimate thromboembolic and bleeding risk, have been specifically validated in oncological patients. A CHA2DS2VASc score ≥ 2 in men and ≥ 3 in women is anyway considered valid to start anticoagulation although tailoring this decision according to the characteristics of each oncological patient is highly recommended (Zamorano et al. 2016). Also, each score should be frequently reassessed considering that comorbidities often become evident shortly after atrial fibrillation has been diagnosed (Chao et al. 2019).

  The choice regarding the best anticoagulation regimen to use is also not obvious. Vitamin K antagonists are often challenging to manage in oncological patients; drug–drug interaction, impairment of kidney and/or liver function, and altered gastrointestinal absorption can lead to a large spectrum of complications ranging from inefficacy to major bleedings. Low molecular weight heparin can confer non-negligible advantages in safety profile and manageability, but lack of oral route of administration may significantly worsen quality of life (Sanz et al. 2019). Furthermore, long lasting treatment can cause side effects rarely seen in clinical practice such as osteoporosis. Oncological patients, in particular patients with advanced disease, were initially under-represented in randomized clinical trials investigating the safety profile and efficacy of direct oral anticoagulants. However, in the last years, use of direct oral anticoagulants has been progressively endorsed on the basis of data from retrospective studies, registries, and clinical trials. Data are still conflicting in patients suffering from gastrointestinal and urological cancers, in whom it may be cautious to avoid direct oral anticoagulants (Wojtukiewicz et al. 2020). Additional high quality studies are still needed to help guiding clinical practice in this specific field of cardio-oncology (Shah et al. 2018; Lopez Fernandez and Van der Meer 2019; Wojtukiewicz et al. 2020).

5.1.5.2 QTc prolongation and related arrhythmias

Corrected QT interval and electrolytes should be measured at baseline before chemotherapy initiation, 1–2 weeks after initial dose (or variation of the same), then monthly for the first 3 months (with the exception of arsenic trioxide that should be monitored weekly) (Zamorano et al. 2016). If concern exists, QTc should be measured after every chemotherapy cycle and when the used drug reaches steady state (Tamargo et al. 2015). Subsequently, long-term follow-up should be individualized. Closer follow-up is required in case of documented electrolyte derangements or in case of predisposing clinical conditions (e.g., diarrhea) (Zamorano et al. 2016). Once relevant QTc prolongation is detected (QTc > 500 ms or ΔQT > 60 ms, but some exceptions exist such as nilotinib, with an upper QTc cutoff of 480 ms) daily monitoring is warranted. Discontinuation of co-administered drugs with known QTc prolonging effects (such as anti-emetics, antibiotics, and others that can be found at www.crediblemeds.org) and electrolytes correction are the first therapeutic measures to adopt. Bradycardia should also be corrected, if necessary, with temporary pacing. Offending chemotherapeutic drug should be temporally suspended and resumed once QTc normalizes, at reduced dose if indicated. Balancing potential side effects of chemotherapeutics is essential; given the high lethality of some malignancies, benefits of the drugs could outweigh risk of QTc prolongation. In these cases, monitoring should be even closer. If torsades de points or similar arrhythmias develop (and also after a spontaneous resolution, until the cause is not removed), magnesium sulfate should be administered. Isoprenaline can be useful to increase heart rate and to reduce the period of susceptibility to major ventricular arrhythmias in heart cycle, but it is associated with an arrhythmogenic effect. A similar effect can be obtained by placing a temporary transvenous pacemaker or with re-programming of a pre-existing definitive pacemaker/implantable cardioverter defibrillator to a lower rate limit ≥90 bpm (Priori et al. 2015; Zamorano et al. 2016).

5.1.5.3 AV conduction disturbances and sinus node dysfunction

As mentioned before, taxanes are the drugs most frequently associated with bradycardia which is well tolerated and asymptomatic in the vast majority of cases (Lopez Fernandez and Van der Meer 2019). An attempt to remove causative factors (including comedications with negative chronotropic effect) remains the best strategy. If not feasible, pacemaker implantation (temporary or definitive) is the treatment of choice (Zamorano et al. 2016).

5.1.5.4 Supraventricular arrhythmias

With the exception of atrial fibrillation, as discussed above, treatment of supraventricular arrhythmias does not differ from non-oncological patients (Zamorano et al. 2016). The majority of supraventricular arrhythmias are easily managed with a high percentage of success, but the feasibility and net clinical benefit of invasive treatments (i.e., transcatheter ablation) need to be carefully weighted in such a frail population.

5.1.5.5 Ventricular arrhythmias

Acute management of ventricular arrhythmias is beyond the scope of Chapter V. QTc interval prolongation has been discussed above. In oncological patients, ventricular arrhythmias may be a direct effect of cardiotoxic drugs and/or radiotherapy or an indirect manifestation of induced ischemia or electrolyte derangements. Predisposing factors should be removed whenever possible; discontinuation of chemothcrapeutics should always be carefully weighted taking into account the clinical condition of each patient and the severity of the arrhythmias. Indeed, grade 4 complications (i.e., life-threatening) preclude further use of the offending drugs.

5.1.6. Prophylaxis

As in the general population, also in patients with cancer prevention of arrythmias should start from cardiovascular risk factors control. Every effort to avoid the development of structural heart disease should be implemented.

The possibility to predict which patient will experience a reduction of left ventricular function (with or without heart failure symptoms) remains an ambitious aim. The echocardiographic assessment of global longitudinal strain (GLS) and serial measurements of biomarkers such as natriuretic peptides (i.e., BNP/NTproBNP) seem promising, but no strong recommendations on their routine use can be provided (Zamorano et al. 2016).

Hypertension, occurring in one out of three patients with cancer, is the commonest cardiovascular comorbidity encountered in clinical practice. Rarely, hypertension can predict and indicate the efficacy of therapies targeting angiogenesis, but, even in those circumstances, it has to be promptly treated. Treatment of hypertension may represent the single most effective preventive action for the entire spectrum of cardiac arrhythmias occurring in patients with cancer (Lopez Fernandez and Van der Meer 2019).

General rules for prevention of coronary artery disease apply also in patients with cancer. Healthy diet, smoking cessation, regular exercise, and weight control can prevent coronary artery disease and seem also effective in improving quality of life and mitigating cardiotoxicity (Zamorano et al. 2016).

Electrolyte derangements have been already discussed before. Prevention of this common clinical problem, often with simple expedients such as hydration, can substantially reduce the risk of arrhythmias.

Pharmacological prevention proved effective in reducing the risk of chemotherapy-induced cardiotoxicity. Candesartan showed a modest preventive effect on left ventricular ejection fraction reduction induced by anthracyclines, while perindopril, bisoprolol, and metoprolol failed to show significant effect. Betablockers showed instead a significative preventive effect in reducing the incidence of left ventricular systolic dysfunction and heart failure in patients receiving anthracyclines associated with trastuzumab. Pharmacological preventive strategies seem reasonable when troponin elevation is observed consequently to chemotherapy. Dexrazoxane is approved for the prevention of toxicity induced by doxorubicin and epirubicin when exceeding dose of 300 and 540 mg/m^2, respectively (Zamorano et al. 2016).

One of the most arrhythmogenic situations often encountered in cardio-oncology is surgery. Atrial fibrillation in this setting is particularly challenging. The role of perioperative beta blockade has been a matter of debate for years with recent studies recommending its use before cardiac surgery but not before non-cardiac surgery because an increased risk of death and stroke has been observed (Blessberger et al. 2018). Amiodarone has been proven effective in preventing the occurrence of atrial fibrillation after cardiac surgery, with an additive effect on beta-blockers (Auer et al. 2004). Data are currently insufficient to advise the use of other drugs.

The adoption of techniques oriented to reduce dose absorption by the heart during radiotherapy is obviously beneficial. Modern techniques brought improvement in the prevention of actinic cardiomyopathy; however, irradiation of the heart remains unavoidable when targeting adjacent structures (Zamorano et al. 2016).

In conclusion, prevention of cancer-related arrhythmias has been poorly investigated so far and many knowledge gaps still exist. Every effort should be made to allow patients with cancer to receive needed treatments without being limited by preventable complications.

References

1. Ajero PM, Belsky JL, Prawius HD, et al. Chemotherapy-induced hypocalcemia. Endocr Pract. 2010;16:284–90.
2. Arbuck SG, Strauss H, Rowinsky E, et al. A reassessment of cardiac toxicity associated with Taxol. J Natl Cancer Inst Monogr. 1993:117–30.
3. Auer J, Weber T, Berent R, et al.; Study of Prevention of Postoperative Atrial Fibrillation. A comparison between oral antiarrhythmic drugs in the prevention of atrial fibrillation after cardiac surgery: the pilot study of prevention of postoperative atrial fibrillation (SPPAF), a randomized, placebo-controlled trial. Am Heart J. 2004;147:636–43.
4. Berardi R, Torniai M, Lenci E, et al. Electrolyte disorders in cancer patients: a systematic review. J Cancer Metastasis Treat 2019;5:79.
5. Bischiniotis TS, Lafaras CT, Platogiannis DN, et al. Intrapericardial cisplatin administration after pericardiocentesis in patients with lung adenocarcinoma and malignant cardiac tamponade. Hellenic J Cardiol. 2005;46:324–9.
6. Blessberger H, Kammler J, Domanovits H, et al. Perioperative beta-blockers for preventing surgery-related mortality and morbidity. Cochrane Database Syst Rev. 2018;3:CD004476.
7. Bogossian H, Linz D, Heijman J, et al. QTc evaluation in patients with bundle branch block. Int J Cardiol Heart Vasc. 2020;30:100636.
8. Brell JM. Prolonged QTc interval in cancer therapeutic drug development: defining arrhythmic risk in malignancy. Prog Cardiovasc Dis. 2010;53:164–72.
9. Brugada J, Katritsis DG, Arbelo E, et al.; ESC Scientific Document Group. 2019 ESC Guidelines for the management of patients with supraventricular tachycardiaThe Task Force for the management of patients with supraventricular tachycardia of the European Society of Cardiology (ESC). Eur Heart J. 2020;41:655–720.

10. Buza V, Rajagopalan B, Curtis AB. Cancer Treatment-Induced Arrhythmias: Focus on Chemotherapy and Targeted Therapies. Circ Arrhythm Electrophysiol. 2017;10:e005443.
11. Chang HM, Okwuosa TM, Scarabelli T, et al. Cardiovascular Complications of Cancer Therapy: Best Practices in Diagnosis, Prevention, and Management: Part 2. J Am Coll Cardiol. 2017;70:2552–2565.
12. Chao TF, Liao JN, Tuan TC, et al. Incident Co-Morbidities in Patients with Atrial Fibrillation Initially with a CHA2DS2-VASc Score of 0 (Males) or 1 (Females): Implications for Reassessment of Stroke Risk in Initially 'Low-Risk' Patients. Thromb Haemost. 2019;119:1162–1170.
13. Dhesi S, Chu MP, Blevins G, et al. Cyclophosphamide-Induced Cardiomyopathy: A Case Report, Review, and Recommendations for Management. J Investig Med High Impact Case Rep. 2013;1:2324709613480346.
14. Farmakis D, Parissis J, Filippatos G. Insights into onco-cardiology: atrial fibrillation in cancer. J Am Coll Cardiol. 2014;63:945–53.
15. Fuentes HE, Tafur AJ, Caprini JA. Cancer-associated thrombosis. Dis Mon. 2016;62:121–58.
16. Galetta F, Franzoni F, Cervetti G, et al. Effect of epirubicin-based chemotherapy and dexrazoxane supplementation on QT dispersion in non-Hodgkin lymphoma patients. Biomed Pharmacother. 2005;59:541–4.
17. Giustozzi M, Ali H, Reboldi G, et al. Safety of catheter ablation of atrial fibrillation in cancer survivors. J Interv Card Electrophysiol. 2020. doi: 10.1007/s10840-020-00745-7.
18. Guglin M, Aljayeh M, Saiyad S, et al. Introducing a new entity: chemotherapy-induced arrhythmia. Europace. 2009;11:1579–86.
19. Herrmann J. Adverse cardiac effects of cancer therapies: cardiotoxicity and arrhythmia. Nat Rev Cardiol. 2020;17:474–502.
20. Hindricks G, Potpara T, Dagres N, et al.; ESC Scientific Document Group. 2020 ESC Guidelines for the diagnosis and management of atrial fibrillation developed in collaboration with the European Association for Cardio-Thoracic Surgery (EACTS). Eur Heart J. 2021;42:373–498.
21. Hu YF, Liu CJ, Chang PM, et al. Incident thromboembolism and heart failure associated with new-onset atrial fibrillation in cancer patients. Int J Cardiol. 2013;165:355–7.
22. Kamineni P, Prakasa K, Hasan SP, et al. Cardiotoxicities of paclitaxel in African Americans. J Natl Med Assoc. 2004;96:995.
23. Kirchhof P, Camm AJ, Goette A, et al.; EAST-AFNET 4 Trial Investigators. Early Rhythm-Control Therapy in Patients with Atrial Fibrillation. N Engl J Med. 2020;383:1305–1316

24. Lipsky AH, Farooqui MZ, Tian X, et al. Incidence and risk factors of bleeding-related adverse events in patients with chronic lymphocytic leukemia treated with ibrutinib. Haematologica. 2015;100:1571–8
25. Lopez Fernandez T and Van der Meer P. Cardio-oncology: it is not only heart failure! e Journal of Cardiology Practice 2019; 16: 38.
26. Mir H, Alhussein M, Alrashidi S, et al. Cardiac Complications Associated With Checkpoint Inhibition: A Systematic Review of the Literature in an Important Emerging Area. Can J Cardiol. 2018;34:1059–1068.
27. Muluneh B, Richardson DR, Hicks C, et al. Trials and Tribulations of Corrected QT Interval Monitoring in Oncology: Rationale for a Practice-Changing Standardized Approach. J Clin Oncol. 2019;37:2719–2721.
28. Naing A, Veasey-Rodrigues H, Hong DS, et al. Electrocardiograms (ECGs) in phase I anticancer drug development: the MD Anderson Cancer Center experience with 8518 ECGs. Ann Oncol. 2012;23:2960–2963.
29. Natasha G, Chan M, Gue YX, et al. Fatal heart block from intentional yew tree (Taxus baccata) ingestion: a case report. Eur Heart J Case Rep. 2019;4:1–4.
30. Osuna PM, Udovcic M, Sharma MD. Hyperthyroidism and the Heart. Methodist Debakey Cardiovasc J. 2017;13:60–63.
31. Porta-Sánchez A, Gilbert C, Spears D, et al. Incidence, Diagnosis, and Management of QT Prolongation Induced by Cancer Therapies: A Systematic Review. J Am Heart Assoc. 2017;6:e007724.
32. Prandoni P, Lensing AW, Piccioli A, et al. Recurrent venous thromboembolism and bleeding complications during anticoagulant treatment in patients with cancer and venous thrombosis. Blood. 2002;100:3484–8.
33. Priori SG, Blomström-Lundqvist C, Mazzanti A, et al.; ESC Scientific Document Group. 2015 ESC Guidelines for the management of patients with ventricular arrhythmias and the prevention of sudden cardiac death: The Task Force for the Management of Patients with Ventricular Arrhythmias and the Prevention of Sudden Cardiac Dcath of the European Society of Cardiology (ESC). Endorsed by: Association for European Paediatric and Congenital Cardiology (AEPC). Eur Heart J. 2015;36:2793–2867.
34. Richards WG, Zellos L, Bueno R, et al. Phase I to II study of pleurectomy/decortication and intraoperative intracavitary hyperthermic cisplatin lavage for mesothelioma. J Clin Oncol. 2006;24:1561–7.
35. Sanz AP, Gómez JLZ. AF in Cancer Patients: A Different Need for Anticoagulation? Eur Cardiol. 2019;14:65–67

36. Shah S, Norby FL, Datta YH, et al. Comparative effectiveness of direct oral anticoagulants and warfarin in patients with cancer and atrial fibrillation. Blood Adv. 2018;2:200–209.
37. Siegel D, Martin T, Nooka A, et al. Integrated safety profile of single-agent carfilzomib: experience from 526 patients enrolled in 4 phase II clinical studies. Haematologica. 2013;98:1753–61.
38. Sury K, Perazella MA, Shirali AC. Cardiorenal complications of immune checkpoint inhibitors. Nat Rev Nephrol. 2018;14:571–588. +
39. Tamargo J, Caballero R, Delpón E. Cancer chemotherapy and cardiac arrhythmias: a review. Drug Saf. 2015;38:129–52.
40. Tomkowski WZ, Filipecki S. Intrapericardial cisplatin for the management of patients with large malignant pericardial effusion in the course of the lung cancer. Lung Cancer. 1997;16:215–22.
41. Van Gelder IC, Groenveld HF, Crijns HJ, et al.; RACE II Investigators. Lenient versus strict rate control in patients with atrial fibrillation. N Engl J Med. 2010;362:1363–73.
42. Vandenberk B, Vandael E, Robyns T, et al. Which QT Correction Formulae to Use for QT Monitoring? J Am Heart Assoc. 2016;5:e003264.
43. Wojtukiewicz MZ, Skalij P, Tokajuk P, et al. Direct Oral Anticoagulants in Cancer Patients. Time for a Change in Paradigm. Cancers (Basel). 2020;12:11.
44. Zamorano JL, Lancellotti P, Rodriguez Muñoz D, et al.; ESC Scientific Document Group. 2016 ESC Position Paper on cancer treatments and cardiovascular toxicity developed under the auspices of the ESC Committee for Practice Guidelines: The Task Force for cancer treatments and cardiovascular toxicity of the European Society of Cardiology (ESC). Eur Heart J. 2016;37:2768–2801.

6

Cancer in the Heart Failure Population

Alessandra Cuomo, MD[1]; Flora Pirozzi, MD, PhD[1]; Francesca Paudice, MD[1]; Giovanni Perrotta, MD[1]; Giovanni D'Angelo, MD[1]; Antonio Carannante, MD[1]; Carlo Gabriele Tocchetti, MD, PhD, FHFA, FISC[1,2,3]; Valentina Mercurio, MD, PhD, FISC[1]; Pietro Ameri MD, PhD, FHFA[4,5]

[1]Department of Translational Medical Sciences, Federico II University, Naples, Italy
[2]Interdepartmental Center of Clinical and Translational Research (CIRCET), Federico II University, Naples, Italy
[3]Interdepartmental Hypertension Research Center (CIRIAPA), Federico II University, Naples, Italy
[4]Cardiovascular Disease Unit, IRCCS Italian Cardiovascular Network, IRCCS Ospedale Policlinico San Martino, Genoa, Italy
[5]Department of Internal Medicine, University of Genova, Genoa, Italy

Correspondence to:
Alessandra Cuomo, MD
Dipartimento di Scienze Mediche Traslazionali
Università degli Studi di Napoli Federico II
Via Sergio Pansini, 5
80131 Naples, Italy
Phone. +39-081-746-2242
Fax: +39-081-746-2246
Email: alebcuomo@gmail.com

KEYWORDS: Cancer; Cardiotoxicity; Cardiovascular Diseases;Heart Failure; Preventive Cardiology;

6.1 Introduction

Along with the increase in life expectancy, the incidence of conditions associated with aging, such as cancer and cardiovascular diseases (CVDs), has also increased, and this trend is expected to persist over the next decade (Heidenreich et al. 2011). Heart failure (HF) represents the final stage of different CVDs and its incidence has notably grown during last decades (Ponikowski et al. 2016; Benjamin et al. 2018). Furthermore, it is common knowledge that cancer and cardiovascular diseases are responsible for most of non-accidental deaths in industrialized countries and listed among the leading causes of mortality in the world (Ameri et al. 2018; Anker et al. 2018).

At first, Cardio-Oncology focused predominantly on the development of cardiac toxicity due to oncological treatments, such as anthracyclines, well-known for their adverse cardiovascular effects,with several antineoplastic treatments being associated with cardiovascular side effects that may lead to new-onset HF (Zamorano et al. 2016; Armenian et al. 2017; Denlinger et al. 2018). Over the past decade, the field of cardio-oncology has been expanding, acquiring more importance in the management of cancer patients, considering the tight link between cancer and CVDs (Lenihan et al. 2016; Moslehi et al. 2019). Indeed, not only HF can be induced by cancer therapies, but recent research has pointed out that the presence of HF itself might promote tumorigenesis (de Boer et al. 2019). In this setting, the management of HF patients who develop cancer represents a challenge for cardio-oncologists (Aboumsallem et al. 2020), considering that little is known about standard of care of patients with pre-existing HF who develop malignancies. Furthermore, most trials on antineoplastic treatments exclude patients with HF because of their numerous comorbidities and worse prognosis, compared to the general population (Ameri et al. 2018). Aim of Chapter VI is to discuss the epidemiology of new-onset cancer in the HF population, the pathophysiological link between those two diseases and to explore possible clinical implication for practitioners. Figure 6.1 summarizes the links between HF and cancer, by means of epidemiology, phatophysiology and clinical implications.

6.2 Heart Failure and New Onset Cancer: Epidemiology

The progressive aging of world population predisposes to an increased incidence of HF, which represents a primary cause of morbidity and mortality in the world, due to age-related cardiac structure alteration. Moreover, the

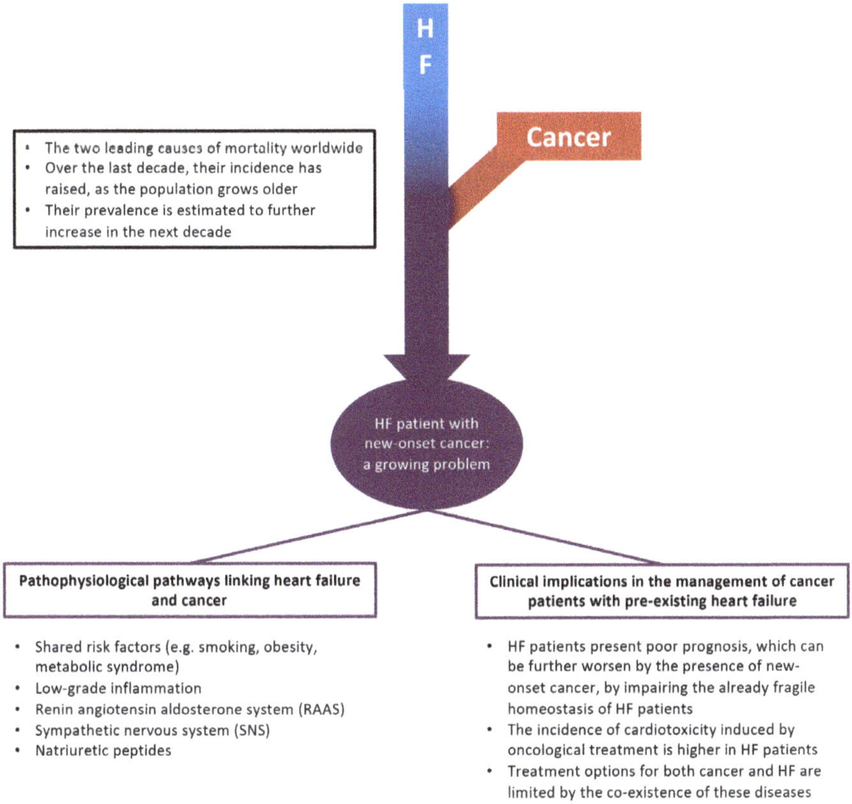

Figure 6.1 Graphic summary of the link between heart failure and cancer

improved treatments of cardiovascular diseases (e.g., acute myocardial infarction, AMI), leading to significant reduction in short-term mortality from these causes, did not affect cardiac remodeling and subsequent development of HF (Hasin et al. 2013; Ameri et al. 2018). Surprisingly, higher mortality and morbidity among HF patients, compared to general population, are more frequently attributed to non-cardiovascular causes than cardiovascular ones (Hasin et al. 2013).

Cancer is another major cause of mortality and its incidence increases with age as well (Hasin et al. 2017). Albeit HF and cancer have been considered as distinguished conditions for a long time, recent evidence shows that they are related and may be considered as comorbidities (Brancaccio et al. 2020). It is wellknown that the use of cardiotoxic oncologic therapies can determine cardiac damages which ultimately may lead to HF (Zamorano et al. 2016).

Besides, the tumor itself can stimulate myocardial dysfunction through systemic alterations (Hasin et al. 2016). Interestingly, some epidemiological and experimental studies have proved that exists a greater risk of cancer development in patients with pre-existing diagnosis of HF (Hasin et al. 2013, 2016, 2017; Banke et al. 2016).

A case-control study explored cancer history among subjects with a new diagnosis of HF (cases) and without HF (controls), finding no association between cancer diagnosis and succeeding development of HF. Then, in a cohort study, the authors examined the long-term risk of developing cancer among HF patients compared to controls, excluding subjects with a prior diagnosis of malignancies, and adjusting for major shared risk factors (e.g., body mass index, smoking, diabetes mellitus, and hypertension). The results showed a 60% increased risk of incident cancer in HF patients, with no sex differences in the association but a 56% higher risk of death in HF patients who developed cancer compared to HF patients who did not (Hasin et al. 2013). In a different prospective cohort study, the same group , investigated the risk of cancer in subjects surviving first acute myocardial infarction (AMI) comparing patients who developed HF after AMI to those who did not develop it. HF patients had a 71% greater risk of later cancer diagnosis. Since these groups had history of AMI, both shared many risk factors and medications (post-AMI treatment); therefore, this study, removing these variables, offers an even stronger evidence of the association between HF (especially with reduced ejection fraction, HFrEF) and cancer (Hasin et al. 2016).

Another cohort study explored the incidence of all types of cancer in a group of Danish patients with HF compared with the general population, and they observed, adjusting for shared risk factors, higher risk of all major types of tumors, except for prostate neoplasms, in HF patients. This study also reported increased mortality among HF patients with new-onset cancer than oncological patients without HF (Banke et al. 2016).

However, it is important to consider that the association between HF and cancer found in these studies can be partially explained by surveillance bias. Patients enrolled in these studies undergo an active follow-up with regular visits and diagnostic tests that may reveal cancer earlier than in general population (e.g., chest X-ray evidence a lung lesion) or discover tumors that would have not been discovered otherwise (Ameri et al. 2018). Furthermore, typical medications used for HF treatment may reveal malignancies that would have been otherwise asymptomatic: for example, the use of anticoagulant and antiplatelets could determine bleeding in patients with unrevealed gastrointestinal neoplasm. Conversely, many symptoms can manifest themselves both in HF and cancer (e.g., dyspnea, fatigue, weight loss, etc.) and could be considered as caused by the advancing of HF rather than the presence of new-onset malignancies, delaying cancer diagnosis (Ameri et al. 2018).

Over the past decade, major improvements in cardiovascular care have led to a significant reduction of the number of deaths due to cardiovascular causes. Consequently, HF patients, whose life expectancy has grown compared to the past, might develop other comorbidities, especially cancer. In a recent paper, Dr. Tini *et al.* (2020) demonstrated that cancer is among the most frequent causes of death in patients with HFrEF. In particular, the authors, performing a meta-analysis on phase 3 randomized controlled trials enrolling patients with HF, found that cancer mortality was not influenced by treatment contrary to what happened for cardiovascular mortality (Tini et al. 2020). Moreover, a subanalysis of the population enrolled in the GISSI-HF trial showed that 3.7% of patients in the trial presented cancer at enrollment and the presence of malignancies increased the risk of death by all causes, after adjusting for age and confounders. Furthermore, among patients who did not present cancer at enrollment, 10.4% died of cancer during follow-up. Finally, patients with cancer-related deaths presented worse NYHA functional class, higher systolic cardiac function, were administered with lower doses of diuretics, showed lower levels of creatinine and uric acid but higher concentrations of cholesterol and hemoglobin, and were characterized by shortened history of HF and better cardiac systolic function (Ameri et al. 2020). Intriguingly, these findings suggest that cancer-related death in the HF population is independent by the severity of HF itself, confirming the hypothesis that cancer significantly worsens HF prognosis (Ameri et al. 2020).

These discoveries suggest the importance of a holistic approach to HF patients and the need to enhance knowledge on the pathophysiological mechanisms underlying the association between HF and cancer in order to define new standards of care for these patients.

6.3 Mechanisms of Cancer Development in the Heart Failure Population

Over the past decades, research has been focusing on unraveling the possible link between cancer and cardiovascular diseases, especially HF. First of all, malignancies and HF share common risk factors like smoking, aging, and metabolic syndrome (MetS) and its components, such as insulin resistance and obesity (de Boer et al. 2019). Furthermore, several mechanisms underlying HF development and persistence are involved in carcinogenesis and cancer spread, including the low-grade inflammatory state typical of HF which may represent a substrate for cancer development in the HF population (van't Klooster et al. 2019).

The presence of MetS, based on the co-existence of at least three conditions among increased fasting glucose, dyslipidemia, central obesity, and systemic hypertension, is itself an important risk factor for the development of cardiovascular diseases (Eckel et al. 2005). Moreover, dyslipidemia seems associated with increased risk of developing colon-rectal cancers (Yao and Tian 2015), while hyperglycemia has been related with higher risk of pancreas, endometrium, and urinary tract cancers, and of malignant melanoma (Stattin et al. 2007). On the other hand, obesity has been associated with increased risk of carcinogenesis (Lauby-Secretan et al. 2016), higher rate of cancer recurrence and recrudescence (Ecker et al. 2019), and worse prognosis in patients already diagnosed with malignancies (Pajares et al. 2013).

Aging itself is characterized by the presence of oxidative stress and cellular senescence, both part of degenerative mechanisms already linked to the development of both cancer and cardiovascular diseases (Abete et al. 1999; Olinski et al. 2007; Liguori et al. 2018; Liberale et al. 2020).

Tobacco use is also an important risk factor for the development of several cardiovascular diseases, such as stroke (Boehme et al. 2017) and systemic hypertension (Virdis et al. 2010), and it has been recognized as an independent risk factor for the development of acute coronary artery disease (Benjamin et al. 2018). On the other hand, tobacco use is well known for its carcinogenic potential and it is recognized as one of the main causes of mortality due to malignancies. In particular, almost 4% of all malignancies in women and 25% in men seem to be smoke-related, while, considering both sexes together, 16% of cancers in industrialized countries and 10% in less developed countries might be attributable to tobacco use (Sasco et al. 2004).

However, as mentioned above, there are other pathways that seem to be involved in both HF and cancer, besides common risk factors. For instance, the renin-angiotensin-aldosterone system (RAAS), the hyperactivation of the sympathetic nervous system (SNS), and the natriuretic peptide system are not only considered as HF hallmarks, but it has also been speculated that they might be associated with increased risk of developing cancer (Sakamoto et al. 2017; Bertero et al. 2019). Indeed, the activation of the RAAS has been proven to be strongly related to cancer spread and neoangiogenesis, through increased expression of angiogenic factors, contributing altogether to worsen cancer prognosis (George et al. 2010). Considering this plausible link between RAAS and malignancies, over the past decade, researches have explored whether RAAS blockers might play a role as antineoplastic drugs, but the results are still controversial (Sipahi et al. 2010; Wang et al. 2013). Intriguingly, in a large meta-analysis, RAAS blockers proved to be effective on all cancer-related endpoints (Sun et al. 2017). A positive correlation between the use of angiotensin converting enzyme inhibitors (ACEi) or angiotensin receptor blockers (ARBs) and cancer risk has been explored, based on the results from the CHARM (Candesartan in Heart Failure: Assessment of Mortality and Morbidity) trial (Pfeffer et al. 2003) and SOLVD (Studies of Left Ventricular Dysfunction) trial. In particular, in the CHARM trial, on candesartan versus placebo, more patients died from cancer in the candesartan group, despite the incidence of non-fatal cancer being similar between both groups (Pfeffer et al. 2003). In the SOLVD trial, on enalapril versus placebo, 43 patients in the enalapril group and 41 patients in the placebo group developed cancer, mostly affecting the gastrointestinal tract, the liver, gallbladder, or pancreas (SOLVD Investigators et al. 1992). Nevertheless, there might be an important bias to be considered: HF patients receiving life-saving therapies and enrolled in large studies like the CHARM and the SOLVD are admitted more often for outpatient follow-ups and this might contribute to diagnose malignancies that might have gone undiagnosed in the real world (Ameri et al. 2018).

It has also been demonstrated that the angiotensin II pathway has a pivotal role in both angiogenesis via VEGF pathway and carcinogenesis, including cell proliferation and migration. Intriguingly, recent data suggest that angiotensin inhibitors might be included in some antineoplastic protocols for metastatic renal cell carcinoma (McDermott et al. 2015) or might play a relevant role in the neoadjuvant therapy of locally advanced pancreatic cancer in association with FOLFIRINOX, the actual standard of care (Murphy et al. 2019).

Considering the SNS, it is well known that β-adrenergic receptors (βARs) have a central role in HF developing, and their role has been investigated also in other diseases, including tumorigenesis. In particular, it has been hypostasized that βARs hyperactivation may favor carcinogenesis, via the β-arrestin 1 pathway (Hara et al. 2011), and may be involved in cellular proliferation, through the activation of CREB, NF-kB, and AP-1 (Zhang et al. 2010). Furthermore, it seems that the bARs pathway plays an important role in cellular apoptosis, leading to tumor cells resistance through the activation of different mechanisms, such as the inhibition of proapoptotic protein BAD (Hassan et al. 2013), gene suppressor p53 (Zhang et al. 2010), and anoikis (Sood et al. 2010).

Intriguingly, both β1 and β2 adrenergic receptors are widely expressed in all solid cancer cells, suggesting that they might play an important role in malignancies development and tumor cells survival, besides their well-studied and pivotal role in heart function (Barron et al. 2011; Coelho et al. 2017). These data support the hypothesis that β-blockers in cancer patients might be used not only for their cardio-protective effects against cardiotoxicity induced by antineoplastic drugs but also adjuvant drugs in oncological protocols (Sysa-Shah et al. 2016; Avila et al. 2018; Guglin et al. 2019).

Results from different studies suggest that the SNS might also contribute to the establishment of the microenvironment which favors cancer development and growth (Cole et al. 2015). In particular, βARs activation enhances the production of prostaglandin E2 and VEGF-C by tumor-associated macrophages (Galdiero et al. 2013), which ultimately lead to increased density of both lymphoid and blood vessels peri- and intra-neoplastic, favorizing tumor spread and metastasis (Armaiz-Pena et al. 2015).

Finally, βARs seem to be able to suppress the natural killer cells activity, creating an environment favorable for cancer development and dissemination (Shakhar and Ben-Eliyahu 1998).

As stated above, natriuretic peptides also seem to be involved in carcinogenesis, and their role has been recently explored (Kong et al. 2008; Nojiri et al. 2015). It is well known that natriuretic peptides, such as atrial natriuretic peptide (ANP), brain natriuretic peptide (BNP), and its inactive N-terminal portion (NT-proBNP) not only increase during HF but also have a pivotal pathophysiological role (Wong et al. 2017). Interestingly, recent data suggest that the presence of circulating natriuretic peptides, i.e., NT-proBNP, might be involved in tumor progression and severity. These findings also support the hypothesis that the cancer patients may present subclinical morphological or functional heart damage, opening the road to the possible use of HF medications in oncological patients as part of the anticancer protocols, beyond their cardioprotective role against cardiotoxicity induced by chemotherapy (Pavo et al. 2015).

The presence of a mild chronic inflammatory state distinguishes both HF and cancer. Over the past century, the link between inflammation and cancer (Lesterhuis et al. 2011) and the inflammatory state and cardiovascular diseases (van't Klooster et al. 2019) has been extensively explored. Indeed, it has been demonstrated that inflammation plays a pivotal role in the establishment and advancement of the atherosclerotic process (Libby et al. 2019), promoting thrombosis and ultimately leading to the development of ischemic heart disease (Koene et al. 2016). Furthermore, data suggest that HF patients present increased plasma concentrations of pro-inflammatory cytokines (Levine et al. 1990; Testa et al. 1996; Torre-Amione et al. 1996), compatible with the mild chronic inflammatory state that characterizes these patients (Suthahar et al. 2017).

It has also been demonstrated that cells post-myocardial infarction present an intense response to stress, characterized by the activation of NF-kB (Hasin et al. 2013), known to be one of the major promoters of cancer development and growth, leading to the activation of numerous genes involved in different tumor mechanisms, such as cell proliferation, survival, spreading, and angiogenesis (Chaturvedi et al. 2011; Meijers and De Boer 2019).

Concerning pro-inflammatory cytokines, interleukin-1 (IL-1) seems to play a central role in both cardiovascular diseases and cancer. In particular, the Canakinumab Anti-Inflammatory Thrombosis Outcome (CANTOS) trial demonstrated that canakinumab, an antibody targeting IL-1β, was able to reduce the incidence of major cardiovascular events in patients with clinical history of myocardial infarction (Ridker et al. 2017a) and also showed to be effective in reducing the incidence of lung cancer in this setting (Ridker et al. 2017b).

To further support the hypothesis that HF itself is able to promote carcinogenesis, an elegant work from Meijers and colleagues (2018) showed that the failing heart releases factors which promote cancer growth, independently from the hemodynamic impairment (Meijers et al. 2018). Moreover, Meijers *et al.* demonstrated that plasma samples from 101 patients with chronic HF present increased levels of five proteins, compared to 180 healthy patients (Hillege et al. 2001; Schroten et al. 2013) Finally, it was also demonstrated that HF biomarkers and proteins related to inflammation were able to predict the incidence of cancer, independently from the tumor risk factor (Kitsis et al. 2018).

6.4 Cancer in Heart Failure Patients: Clinical Implications

The co-existence of cancer and cardiovascular diseases always represents a challenge for clinicians. For instance, cardio-oncology first aimed at treating cardiotoxicity and HF induced by antineoplastic treatments. Nowadays, physicians not only have to deal with the cardiovascular effects of oncological treatments but also have to relate to HF patients who develop new-onset cancer and need to be treated for both diseases. Indeed, new-onset cancer may present with symptoms that may overlap with those of the pre-existing HF. Furthermore, the presence of new-onset cancer in HF patients worsens the already poor prognosis of these patients and clinicians need to be particularly careful when deciding the optimal treatments for both diseases (Ameri et al. 2018).

Physicians need to be aware that there are numerous clinical implications in this setting: in HF patients, the presence of new-onset cancer impairs the already fragile homeostasis, and it may also increase the risk of developing cardiotoxicity induced by anticancer treatments and can be a burden to both cardiological and oncological therapeutic approaches, leading to a poorer prognosis in this subset of patients. Additionally, new-onset cancer itself may contribute to further compromise heart function (Musolino et al. 2019), for example, inducing electrolytes or hormonal alterations, deteriorating the already compromised endothelium or worsening the chronic inflammation state (Ameri et al. 2018).

A tight collaboration between cardiologists and oncologists is fundamental to ensure the best assistance to all patients with pre-existing HF who develop new-onset cancer. Cardiologists and HF specialists need to perform a comprehensive evaluation of HF patients, not only to diagnose possible new-onset cancers but also to fully assess patients' characteristics. In particular, it is highly suggested to perform a full baseline evaluation, including clinical and family history, physical examination, ECG, blood withdrawal (in order to verify electrolyte status, exclude the presence of anemia, new-onset diabetes, or hormonal alterations), echocardiography, and other tests that are considered necessary by the clinician according to each patient's condition. Moreover, it is essential to perform a risk/benefit analysis for each subject and evaluate all the possible therapeutic options for both HF and cancer in order to choose the best possible treatment management, tailored to each patient's specific needs (Ameri et al. 2018; Lancellotti et al. 2018; Pareek et al. 2018). To provide the best clinical care, the clinician should first address the modifiable comorbidities (i.e., suggest weight loss, when necessary, and smoking cessation) and optimize cardiac therapy, including up-titration of β-blockers, ACEi, ARBs, neprilysin inhibitors, and diuretics and optimization of anti-diabetes medications (Ponikowski et al. 2016; Ameri et al. 2018). Patients should also be evaluated for the presence of pre-existing valvular defects and residual ischemia, addressing these issues when needed. Moreover, it is fundamental to promptly address any hormonal, metabolic, or electrolytes disorder, considering that their persistence may increase patients' risk of developing cardiovascular toxicities induced by oncological treatments (Zamorano et al. 2016; Armenian et al. 2017; Ameri et al. 2018; Denlinger et al. 2018).

Unfortunately, optimization and up-titration of HF specific treatment may be a long process, sometimes requiring months, which could delay the start of oncological treatments, both surgical and chemotherapy, further compromising patients' prognosis (Ameri et al. 2018). For example, patients may require a coronary angioplasty before oncological treatments initiation, leading to the administration of double antiplatelet therapy (DAPT) for at least a month after the angioplasty that can postpone cancer surgery due to the increased risk of bleeding induced by the cardiological treatment. Obviously, physicians need to perform a careful evaluation of the bleeding risk induced by the DAPT on one side and, on the other side, of the risk of intra-stent thrombosis associated with the DAPT suspension (Valgimigli et al., 2018; Neumann et al. 2019).

Clinicians should schedule regular follow-ups for HF-cancer patients to promptly identify any signs or symptoms of clinical worsening, which may be the manifestation of new-onset cardiotoxicity, and to intervene immediately in case of serum alteration or any other signs of decompensate HF. Most importantly, the clinician should recommend withdrawal from oncological treatments only when strictly necessary in order to guarantee to as many HF patients as possible to complete the antineoplastic protocols needed.

Furthermore, cardiologists who treat HF patients who develop cancer need to be fully aware of the fragility of these patients. Hasin and colleagues showed that HF patients who develop cancer are characterized by worse prognosis when compared to patients without HF (Hasin et al. 2013). However, as the authors themselves explain, this data might be related to higher prevalence of comorbidities in the HF population and by the increased mortality risk due to the co-existence of HF and cancer (Mamas et al. 2017).

Additionally, HF patients with malignancies present higher risk of hospitalization, which further compromise their prognosis, increasing the mortality risk (Omersa et al. 2016). As described above, Banke and colleagues explore the incidence of new-onset cancer in the HF population comparing data with the Denmark general background population. In their study, the authors stratified their sample into three subgroups according to age: patients <60 years old, patients aged 60–69, and patients >70 years old. The subanalysis showed that HF patients <60 and 60–69 years old presented the same mortality risk of the general population in the groups aged 60–69 and >70 years old, respectively (Banke et al. 2016). These findings are consistent with the hypothesis that death risk of pre-existing HF overlays is derived by new-onset malignancies, contributing to worsen patients' outcomes.

On the other hand, pre-existing HF and new-onset cancer might be linked by common symptoms, and cancer can further impair the already precarious homeostasis of HF patients (Ameri et al. 2018; Musolino et al. 2019).

Another important issue to be addressed when dealing with both HF patients and cancer patients is the psychological burden, considering that both diseases are associated with depression, which can further negatively impact prognosis (Newhouse and Jiang 2014; Sotelo et al. 2014). Obviously, the psychological impact of the new-onset cancer diagnosis in a patient that is already facing progressive chronic diseases, such as pre-existing HF, can additionally worsen patients' mental status.

Physicians that treat HF patients need also to be aware of the complexity of antineoplastic protocols of cancer patients. In particular, HF patients with new-onset malignancies present higher risk of developing cardiovascular toxicities due to the oncological treatments (Zamorano et al. 2016; Armenian et al. 2017; Ameri et al. 2018; Denlinger et al. 2018). Another issue to be dealt with is the higher perioperative mortality risk of patients with pre-existing HF that can develop cancer that can limit oncological treatments and further worsen prognosis (Smit-Fun and Buhre 2016; Kravchenko et al. 2015).

Another important issue to be faced is the management of fluids during chemotherapy administration. Indeed, many antineoplastic treatments present renal toxicity among their adverse effects and, for this reason, are administered with large amount of fluids to be diluted (Cosmai et al. 2016). Needless to underline that HF patients often cannot be administered with as many fluids as standard patients, considering the risk of fluids overload. On the other hand, HF patients often present chronic kidney failure among their comorbidities, thus increasing the risk of renal toxicity due to anticancer treatments (Schefold et al. 2016). To reduce the risk of renal impairment, fluid doses should be reduced in HF patients, while the time of infusion could be prolonged, and diuretic therapy should be tailored to patient's need, for example, increased when large amount of fluids have to be administered, to avoid pulmonary edema and other manifestations of fluids overload (Ponikowski et al. 2016; Ameri et al. 2018).

Cancer itself is also a predisposing factor for arrhythmias, such as QT prolongation and atrial fibrillation (AFib; Farmakis et al. 2014), which has been related to increased mortality risk in the HF population (Mamas et al. 2009). Although the pathophysiological mechanisms are not quite understood, it seems that AFib in cancer patients might be due to thoracic surgery, advanced age, metabolic or electrolyte impairment, and hypoxia (Farmakis et al. 2014). Recent data suggest that oral anticoagulants are superior to warfarin, thanks to their safety profile and less drug–drug interaction (Park and Khorana 2019). Indeed, it is fundamental to periodically interrogate patients' electronic devices when present, such as pacemakers and implantable defibrillators. High attention should be paid to patients who undergo radiotherapy, considering that radiation may impair the functionality of such electronic devices (Viganego et al. 2016).

When dealing with HF-cancer patients, clinicians need to remember that cancer itself might have both pro-hemorrhagic and prothrombosis effects, according to tumor characteristics, and the management of DAPT and anticoagulation treatment has to be tailored on each specific case. For example, the presence gastrointestinal malignancies or central nervous system metastasis are associated with higher risk of major bleeding and anticoagulation treatment should be avoided, when possible (Angelini et al. 2019). For example, in patients with Afib and colon-rectal tumors, left atrium appendage occlusion might be considered to avoid the bleeding risk associated with anticoagulation treatment in this peculiar subset of patients (Meier et al. 2014; Glikson et al. 2020; Hindricks et al. 2020). On the other hand, thrombotic diathesis is a common feature in cancer patients who, thus, are more likely to develop deep vein thrombosis, pulmonary embolisms, and central venous catheter thrombosis (Park and Khorana 2019).

Hence, a correct bleeding/thrombosis risk stratification is pivotal in HF patients who develop cancer in order to identify if anticoagulation treatment is strictly necessary and which anticoagulant treatment is more appropriate for each patient, choosing between low-weight heparin and oral anticoagulants, according also to possible drug–drug interactions (Kraaijpoel et al. 2018).

Invasive HF treatments are questioned in the HF-cancer population, such as *ex novo* electronic device implantation in patients with cardiac failure with diagnosis of new-onset malignancies (Singh et al. 2019), considering that implantation is vetoed with less than 1 year life expectancy (Ponikowski et al. 2016). For HF-cancer patients with life expectancy of 2 or more years, a valid alternative could be treatment with left ventricular assistant devices (Viganego et al. 2016; Ameri et al. 2018).

Considering that the presence of cancer is an exclusion criterion for the heart transplant list, cardiologists and oncologists should also meticulously assess every HF-cancer patient in need for heart transplantation (Meijers and Moslehi 2019), scrupulously evaluating life expectancy and prognosis for each individual (Ponikowski et al. 2016; Ameri et al. 2018). On the other hand, after the heart transplant, patients need to be administered with life-long treatment with immunosuppressant drugs, well-known for increasing the risk of developing new-onset cancer. However, it is not quite understood if patients with recent diagnosis of malignancies are more prone to develop new caner compared to the general population when administered with immunosuppressors.

As stated above, HF itself usually is among the exclusion criteria for oncological trials, leading to scarce information on the management of many antineoplastic protocols in patients with cardiac conditions, including dosage adjustment and adverse reactions. Recently, the SAFE-HEaRt trial has been investigating the efficacy and safety of administering HER2 inhibitors in patients with mildly reduced HF (Lynce et al. 2017; Angelini et al. 2019). Obviously, more studies are required to define optimal cardioprotective and surveillance strategies to adopt in HF patients who develop cancer and which cardiac patients must withdrawal from oncological treatments due to excessive risks.

In conclusion, HF patients who develop cancer should receive the most appropriate HF therapies to be able to be administered with the best oncological protocol available: neither HF should burden anticancer treatment nor the contrary. Unfortunately, Gross *et al.* (2007) showed that HF patients with colon-rectal malignancies are less likely to be administered with adjuvant oncological protocols and, thus, present worse 5 years prognosis, compared to non-HF patients with the same cancers.

By contrast, HF in patients who are diagnosed with cancer should not be undertreated, considering that the optimized treatment for cardiac failure is composed of a complex set of drugs for both the specific heart disease and for its comorbidities, such as dyslipidemia, dysthyroidism, and diabetes mellitus (Ponikowski et al. 2016). Drug interactions between cardiovascular and oncological drugs should be explored in order to suspend or modify dosages when necessary. Moreover, HF patients' homeostasis might be impaired by the oncological condition, considering that both cancer itself and antineoplastic treatment might induce vomiting, diarrhea, or other endocrinological alterations, leading to electrolyte impairment which has to be promptly addressed by cardio-oncologists. In particular, HF patients with cancer diagnosis might need temporary dosage adjustments and treatment suspension. However, it is important to tailor therapy for each patient, trying to avoid HF specific therapy undertreatment or definitive suspension.

All things considered, each HF patient who develops cancer has to be treated as a unique case, both oncological and HF therapies need to be tailored to the specific individual and clinicians should carefully risk-stratify patients in order to identify those more exposed to the risk of developing new-onset malignancies and provide adequate screening programs.

6.5 Conclusions

Nowadays, it is clear that HF and cancer are not only linked by common pathophysiological mechanisms, but recent data reinforce the hypothesis that cardiac failure itself might promote tumorigenesis. The higher incidence of cancer in HF patients compared to the general population is consistent with those new findings. Moreover, tumor development in patients with pre-existing HF further complexifies both oncological and cardiological management. Not only both cancer and HF are burdened by independent increase in mortality risk but may also interfere with one another's treatment, resulting in a further impairment of patients' prognosis.

A tight collaboration between cardiologists and oncologists is strongly recommended, aimed at improving patients' quality of life and survival. Both professional categories need to rely on one another in order to avoid suboptimization or suspension of either cardiac or antineoplastic treatments. Each patient should be considered as a *unicum* and a multidisciplinary approach, including specialized nurses, cardiac rehabilitation specialists, psychologists, and palliative care professionals, should be always preferred (Meijers and Moslehi 2019).

In the HF population, cancer surveillance programs should be considered to screen those who are more exposed to develop malignancies and intervene as promptly as possible. Nevertheless, more studies on possible cancer biomarkers associated with HF are needed (Kitsis et al. 2018).

Unfortunately, current knowledge being still limited, clinicians need to base most of their clinical decisions on populations other than HF-cancer patients.

References

1. Abete P, Napoli C, Santoro G, et al. Age-related decrease in cardiac tolerance to oxidative stress. J Mol Cell Cardiol. 1999;31:227–36.
2. Aboumsallem JP, Moslehi J, de Boer RA. Reverse Cardio-Oncology: Cancer Development in Patients With Cardiovascular Disease. J Am Heart Assoc. 2020;9:e013754.
3. Ameri P, Canepa M, Anker MS, et al. Cancer diagnosis in patients with heart failure: epidemiology, clinical implications and gaps in knowledge. Eur J Heart Fail. 2018;20:879–887
4. Ameri P, Canepa M, Luigi Nicolosi G, et al.; GISSI-HF Investigators. Cancer in chronic heart failure patients in the GISSI-HF trial. Eur J Clin Invest. 2020;50:e13273.
5. Angelini DE, Radivoyevitch T, McCrae KR, et al. Bleeding incidence and risk factors among cancer patients treated with anticoagulation. Am J Hematol. 2019;94:780–785.
6. Anker MS, von Haehling S, Landmesser U, et al. Cancer and heart failure-more than meets the eye: common risk factors and co-morbidities. Eur J Heart Fail. 2018;20:1382–1384.

7. Armaiz-Pena GN, Gonzalez-Villasana V, Nagaraja AS, et al. Adrenergic regulation of monocyte chemotactic protein 1 leads to enhanced macrophage recruitment and ovarian carcinoma growth. Oncotarget 2015;6:4266–4273.
8. Armenian SH, Lacchetti C, Barac A, et al. Prevention and Monitoring of Cardiac Dysfunction in Survivors of Adult Cancers: American Society of Clinical Oncology Clinical Practice Guideline. J Clin Oncol. 2017;35:893–911.
9. Avila MS, Ayub-Ferreira SM, de Barros Wanderley MR Jr, et al. Carvedilol for Prevention of Chemotherapy-Related Cardiotoxicity: The CECCY Trial. J Am Coll Cardiol. 2018;71:2281–2290.
10. Banke A, Schou M, Videbæk L, et al. Incidence of cancer in patients with chronic heart failure: A long-term follow-up study. Eur J Heart Fail. 2016;18:260–266.
11. Barron TI, Connolly RM, Sharp L, et al. Beta blockers and breast cancer mortality: a population- based study. J Clin Oncol. 2011;29:2635–44.
12. Benjamin EJ, Virani SS, Callaway CW, et al.; American Heart Association Council on Epidemiology and Prevention Statistics Committee and Stroke Statistics Subcommittee. Heart Disease and Stroke Statistics-2018 Update: A Report From the American Heart Association. Circulation. 2018;137:e67–e492.
13. Bertero E, Ameri P, Maack C. Bidirectional Relationship Between Cancer and Heart Failure: Old and New Issues in Cardio-oncology. Card Fail Rev. 2019;5:106–111.
14. Boehme AK, Esenwa C, Elkind MS. Stroke Risk Factors, Genetics, and Prevention. Circ Res. 2017;120:472–495.
15. Brancaccio M, Pirozzi F, Hirsch E, et al. Mechanisms underlying the cross-talk between heart and cancer. J Physiol. 2020; 598:3015–3027.
16. Chaturvedi MM, Sung B, Yadav VR, et al. NF-κB addiction and its role in cancer: 'one size does not fit all'. Oncogene. 2011;30:1615–30.
17. Coelho M, Soares-Silva C, Brandão D, et al. β-Adrenergic modulation of cancer cell proliferation: available evidence and clinical perspectives. J Cancer Res Clin Oncol. 2017;143:275–291.
18. Cole SW, Nagaraja AS, Lutgendorf SK, et al. Sympathetic nervous system regulation of the tumour microenvironment. Nat Rev Cancer. 2015;15:563–72.
19. Cosmai L, Porta C, Gallieni M, et al. Onco-nephrology: A decalogue. Nephrol Dial Transplant 2016;31:515–519.
20. de Boer RA, Meijers WC, van der Meer P, van Veldhuisen DJ. Cancer and heart disease: associations and relations. Eur J Heart Fail. 2019;21:1515–1525.

21. Denlinger CS, Sanft T, Baker KS, et al. Survivorship, Version 2.2018, NCCN Clinical Practice Guidelines in Oncology. J Natl Compr Canc Netw. 2018);16:1216–1247.
22. Eckel RH, Alberti KG, Grundy SM, et al. The metabolic syndrome. Lancet. 2010;375:181 3.
23. Ecker BL, Lee JY, Sterner CJ, et al. Impact of obesity on breast cancer recurrence and minimal residual disease. Breast Cancer Res. 2019;21:1–16.
24. Farmakis D, Parissis J, Filippatos G. Insights into onco-cardiology: Atrial fibrillation in cancer. J Am Coll Cardiol. 2014;63:945–953.
25. Feldman D, Pamboukian SV, Teuteberg JJ, et al.; International Society for Heart and Lung Transplantation. The 2013 International Society for Heart and Lung Transplantation Guidelines for mechanical circulatory support: executive summary. J Heart Lung Transplant. 2013;32: 157–87.
26. Galdiero MR, Garlanda C, Jaillon S, et al. Tumor associated macrophages and neutrophils in tumor progression. J Cell Physiol. 2013; 228:1404–1412.
27. George AJ, Thomas WG, Hannan RD. The renin-angiotensin system and cancer: old dog, new tricks. Nat Rev Cancer. 2010;10:745–59.
28. Glikson M, Wolff R, Hindricks G, et al. EHRA/EAPCI expert consensus statement on catheter-based left atrial appendage occlusion - an update. EuroIntervention. 2020;15:1133–1180.
29. Gross CP, McAvay GJ, Guo Z, et al. The impact of chronic illnesses on the use and effectiveness of adjuvant chemotherapy for colon cancer. Cancer 2007;109:2410–2419.
30. Guglin M, Krischer J, Tamura R, et al. Randomized Trial of Lisinopril Versus Carvedilol to Prevent Trastuzumab Cardiotoxicity in Patients With Breast Cancer. J Am Coll Cardiol. 2019;73:2859–2868.
31. Hara MR, Kovacs JJ, Whalen EJ, et al. A stress response pathway regulates DNA damage through β2-adrenoreceptors and β-arrestin-1. Nature. 2011;477:349–53.
32. Hasin T, Gerber Y, McNallan SM, et al. Patients with heart failure have an increased risk of incident cancer. J Am Coll Cardiol. 2013; 62:881–886.
33. Hasin T, Gerber Y, Weston SA, et al. Heart Failure After Myocardial Infarction Is Associated With Increased Risk of Cancer. J Am Coll Cardiol. 2016; 68:265–271.
34. Hasin T, Iakobishvili Z, Weisz G. Associated Risk of Malignancy in Patients with Cardiovascular Disease: Evidence and Possible Mechanism. Am J Med. 2017;130:780–785.

35. Hassan S, Karpova Y, Baiz D, et al. Behavioral stress accelerates prostate cancer development in mice. J Clin Invest. 2013;123:874–86.
36. Heidenreich PA, Trogdon JG, Khavjou OA, et al. Forecasting the future of cardiovascular disease in the United States: A policy statement from the American Heart Association. Circulation 2011;123:933–944.
37. Hillege HL, Janssen WM, Bak AA, et al.; Prevend Study Group. Microalbuminuria is common, also in a nondiabetic, nonhypertensive population, and an independent indicator of cardiovascular risk factors and cardiovascular morbidity. J Intern Med. 2001;249:519–26.
38. Hindricks G, Potpara T, Dagres N, et al.; ESC Scientific Document Group. 2020 ESC Guidelines for the diagnosis and management of atrial fibrillation developed in collaboration with the European Association for Cardio-Thoracic Surgery (EACTS). Eur Heart J. 2021;42: 373–498.
39. Kitsis RN, Riquelme JA, Lavandero S. Heart Disease and Cancer: Are the Two Killers Colluding? Circulation 2018;138:692–695.
40. Koene RJ, Prizment AE, Blaes A, et al. Shared Risk Factors in Cardiovascular Disease and Cancer. Circulation. 2016;133:1104–14.
41. Kong X, Wang X, Xu W, et al. Natriuretic peptide receptor a as a novel anticancer target. Cancer Res. 2008;68:249–56.
42. Kraaijpoel N, Di Nisio M, Mulder FI, et al. Clinical Impact of Bleeding in Cancer-Associated Venous Thromboembolism: Results from the Hokusai VTE Cancer Study. Thromb Haemost. 2018;118:1439–1449.
43. Kravchenko J, Berry M, Arbeev K, et al. Cardiovascular comorbidities and survival of lung cancer patients: Medicare data based analysis. Lung Cancer 2015;88:85–93.
44. Lancellotti P, Suter TM, López-Fernández T, et al. Cardio-Oncology Services: rationale, organization, and implementation. Eur Heart J. 2019;40:1756–1763.
45. Lauby-Secretan B, Scoccianti C, Loomis D, et al.; International Agency for Research on Cancer Handbook Working Group. Body Fatness and Cancer--Viewpoint of the IARC Working Group. N Engl J Med. 2016;375:794–8.
46. Lenihan DJ, Hartlage G, DeCara J, et al. Cardio-Oncology Training: A Proposal From the International Cardioncology Society and Canadian Cardiac Oncology Network for a New Multidisciplinary Specialty. J Card Fail. 2016;22:465–71.
47. Lesterhuis WJ, Haanen JBAG, Punt CJA. Cancer immunotherapy--revisited. Nat Rev Drug Discov. 2011;10:591–600.

48. Levine B, Kalman J, Mayer L, Fillit HM, Packer M. Elevated circulating levels of tumor necrosis factor in severe chronic heart failure. N Engl J Med. 1990;323:236–41.
49. Libby P, Sidlow R, Lin AE, et al. Clonal Hematopoiesis: Crossroads of Aging, Cardiovascular Disease, and Cancer: JACC Review Topic of the Week. J Am Coll Cardiol. 2019;74:567–577.
50. Liberale L, Montecucco F, Tardif JC, et al. Inflamm-ageing: the role of inflammation in age-dependent cardiovascular disease. Eur Heart J. 2020;41:2974–2982.
51. Liguori I, Russo G, Curcio F, et al. Oxidative stress, aging, and diseases. Clin Interv Aging. 2018;13:757–772.
52. Lynce F, Barac A, Tan MT, et al. SAFE-HEaRt: Rationale and Design of a Pilot Study Investigating Cardiac Safety of HER2 Targeted Therapy in Patients with HER2-Positive Breast Cancer and Reduced Left Ventricular Function. Oncologist. 2017;22:518–525.
53. Mamas MA, Caldwell JC, Chacko S, et al. A meta-analysis of the prognostic significance of atrial fibrillation in chronic heart failure. Eur J Heart Fail. 2009;11:676–683.
54. Mamas MA, Sperrin M, Watson MC, et al. Do patients have worse outcomes in heart failure than in cancer? A primary care-based cohort study with 10-year follow-up in Scotland. Eur J Heart Fail. 2017;19:1095–1104.
55. McDermott DF, Cheng SC, Signoretti S, et al. The high-dose aldesleukin "select" trial: a trial to prospectively validate predictive models of response to treatment in patients with metastatic renal cell carcinoma. Clin Cancer Res. 2015;21:561–8.
56. Meier B, Blaauw Y, Khattab AA, et al.; Document Reviewers. EHRA/EAPCI expert consensus statement on catheter-based left atrial appendage occlusion. Europace. 2014;16:1397–416.
57. Meijers WC, De Boer RA. Common risk factors for heart failure and cancer. Cardiovasc Res. 2019;115:844–853.
58. Meijers WC, Maglione M, Bakker SJL, et al. Heart failure stimulates tumor growth by circulating factors. Circulation 2018;138:678–691.
59. Meijers WC, Moslehi JJ. Need for Multidisciplinary Research and Data-Driven Guidelines for the Cardiovascular Care of Patients With Cancer. JAMA 2019;322:1775–1776.
60. Moslehi J, Fujiwara K, Guzik T. Cardio-oncology: a novel platform for basic and translational cardiovascular investigation driven by clinical need. Cardiovasc Res. 2019;115:819–823.

61. Murphy JE, Wo JY, Ryan DP, et al. Total Neoadjuvant Therapy With FOLFIRINOX in Combination With Losartan Followed by Chemoradiotherapy for Locally Advanced Pancreatic Cancer: A Phase 2 Clinical Trial. JAMA Oncol. 2019;5:1020–1027.
62. Musolino V, Palus S, Latouche C, et al. Cardiac expression of neutrophil gelatinase-associated lipocalin in a model of cancer cachexia-induced cardiomyopathy. ESC Heart Fail. 2019;6:89–97.
63. Neumann FJ, Sousa-Uva M, Ahlsson A, et al. 2018 ESC/EACTS Guidelines on myocardial revascularization. Eur Heart J. 2019;40:87–165.
64. Newhouse A, Jiang W. Heart failure and depression. Heart Fail Clin. 2014;10:295–304.
65. Nojiri T, Hosoda H, Tokudome T, et al. Atrial natriuretic peptide prevents cancer metastasis through vascular endothelial cells. Proc Natl Acad Sci U S A. 2015;112:4086–91. Erratum in: Proc Natl Acad Sci U S A. 2018;115:E7883–E7886.
66. Olinski R, Siomek A, Rozalski R, et al. Oxidative damage to DNA and antioxidant status in aging and age-related diseases. Acta Biochim Pol. 2007;54:11–26.
67. Omersa D, Farkas J, Erzen I, et al. National trends in heart failure hospitalization rates in Slovenia 2004–2012. Eur J Heart Fail. 2016;18:1321–1328.
68. Pajares B, Pollán M, Martín M, et al. Obesity and survival in operable breast cancer patients treated with adjuvant anthracyclines and taxanes according to pathological subtypes: A pooled analysis. Breast Cancer Res. 2013;15:1–14.
69. Pareek N, Cevallos J, Moliner P, et al. Activity and outcomes of a cardio-oncology service in the United Kingdom-a five-year experience. Eur J Heart Fail. 2018;20:1721–1731.
70. Park DY, Khorana AA. Risks and Benefits of Anticoagulation in Cancer and Noncancer Patients. Semin Thromb Hemost. 2019;45:629–637.
71. Pavo N, Raderer M, Hulsmann M, et al. Cardiovascular biomarkers in patients with cancer and their association with all-cause mortality. Heart 2015;101:1874–1880.
72. Pfeffer MA, Swedberg K, Granger CB, et al. Effects of candesartan on mortality and morbidity in patients with chronic heart failure: the CHARM-Overall programme. Lancet 2003;362:759–766.
73. Ponikowski P, Voors AA, Anker SD, et al. 2016 ESC Guidelines for the Diagnosis and Treatment of Acute and Chronic Heart Failure. Rev Esp Cardiol (Engl Ed). 2016;69:1167. Erratum in: Rev Esp Cardiol (Engl Ed). 2017;70:309–310.

74. Ridker PM, Everett BM, Thuren T, et al.; CANTOS Trial Group. Antiinflammatory Therapy with Canakinumab for Atherosclerotic Disease. N Engl J Med. 2017;377:1119–1131.
75. Ridker PM, MacFadyen JG, Thuren T, et al.; CANTOS Trial Group. Effect of interleukin-1β inhibition with canakinumab on incident lung cancer in patients with atherosclerosis: exploratory results from a randomised, double-blind, placebo-controlled trial. Lancet. 2017;390:1833–1842.
76. Sakamoto M, Hasegawa T, Asakura M, et al. Does the pathophysiology of heart failure prime the incidence of cancer? Hypertens Res. 2017;40:831–836.
77. Sasco AJ, Secretan MB, Straif K. Tobacco smoking and cancer: a brief review of recent epidemiological evidence. Lung Cancer. 2004;45S2:S3–9.
78. Schefold JC, Filippatos G, Hasenfuss G, et al. Heart failure and kidney dysfunction: Epidemiology, mechanisms and management. Nat Rev Nephrol. 2016;12:610–623.
79. Schroten NF, Ruifrok WPT, Kleijn L, et al. Short-term vitamin D3 supplementation lowers plasma renin activity in patients with stable chronic heart failure: an open-label, blinded end point, randomized prospective trial (VitD-CHF trial). Am Heart J. 2013;166:357–364.e2.
80. Shakhar G, Ben-Eliyahu S. In vivo beta-adrenergic stimulation suppresses natural killer activity and compromises resistance to tumor metastasis in rats. J Immunol. 1998;160:3251–8.
81. Singh JP, Solomon SD, Fradley MG, et al.; MADIT-CHIC Investigators. Association of Cardiac Resynchronization Therapy With Change in Left Ventricular Ejection Fraction in Patients With Chemotherapy-Induced Cardiomyopathy. JAMA. 2019;322:1799–1805.
82. Sipahi I, Debanne SM, Rowland DY, et al. Angiotensin-receptor blockade and risk of cancer: meta-analysis of randomised controlled trials. Lancet Oncol. 2010;11:627–36.
83. Smit-Fun V, Buhre WF. The patient with chronic heart failure undergoing surgery. Curr Opin Anaesthesiol. 2016;29:391–396.
84. SOLVD Investigators, Yusuf S, Pitt B, et al. Effect of Enalapril on Mortality and the Development of Heart Failure in Asymptomatic Patients with Reduced Left Ventricular Ejection Fractions. N Engl J Med. 1992; 327:685–691.
85. Sood AK, Armaiz-Pena GN, Halder J, et al. Adrenergic modulation of focal adhesion kinase protects human ovarian cancer cells from anoikis. J Clin Invest. 2010;120:1515–23.
86. Sotelo JL, Musselman D, Nemeroff C. The biology of depression in cancer and the relationship between depression and cancer progression. Int Rev Psychiatry 2014;26:16–30.

87. Stattin P, Björ O, Ferrari P, et al. Prospective study of hyperglycemia and cancer risk. Diabetes Care 2007;30:561–567.
88. Sun H, Li T, Zhuang R, et al. Do renin-angiotensin system inhibitors influence the recurrence, metastasis, and survival in cancer patients?: Evidence from a meta-analysis including 55 studies. Medicine (Baltimore). 2017;96:e6394.
89. Suthahar N, Meijers WC, Silljé HHW, et al. From Inflammation to Fibrosis-Molecular and Cellular Mechanisms of Myocardial Tissue Remodelling and Perspectives on Differential Treatment Opportunities. Curr Heart Fail Rep. 2017;14:235–250.
90. Sysa-Shah P, Tocchetti CG, Gupta M, et al. Bidirectional cross-regulation between ErbB2 and β-adrenergic signalling pathways. Cardiovasc Res. 2016;109:358–73.
91. Testa M, Yeh M, Lee P, et al. Circulating levels of cytokines and their endogenous modulators in patients with mild to severe congestive heart failure due to coronary artery disease or hypertension. J Am Coll Cardiol. 1996;28:964–71.
92. Tini G, Bertero E, Signori A, et al. Cancer Mortality in Trials of Heart Failure With Reduced Ejection Fraction: A Systematic Review and Meta-Analysis. J Am Heart Assoc. 2020;9:e016309.
93. Torre-Amione G, Kapadia S, Lee J, et al. Tumor necrosis factor-alpha and tumor necrosis factor receptors in the failing human heart. Circulation. 1996;93:704–11.
94. Valgimigli M, Bueno H, Byrne RA, et al. 2017 ESC focused update on dual antiplatelet therapy in coronary artery disease developed in collaboration with EACTS. Eur J Cardio-thoracic Surg. 2018;53:34–78.
95. van't Klooster CC, Ridker PM, Hjortnaes J, et al. The relation between systemic inflammation and incident cancer in patients with stable cardiovascular disease: a cohort study. Eur Heart J. 2019;3901–3909.
96. Viganego F, Singh R, Fradley MG. Arrhythmias and Other Electrophysiology Issues in Cancer Patients Receiving Chemotherapy or Radiation. Curr Cardiol Rep. 2016;18:52.
97. Virdis A, Giannarelli C, Neves MF, et al. Cigarette smoking and hypertension. Curr Pharm Des. 2010;16:2518–2525.
98. Wang KL, Liu CJ, Chao TF, et al. Long-term use of angiotensin II receptor blockers and risk of cancer: a population-based cohort analysis. Int J Cardiol. 2013;167:2162–6.
99. Wong PCY, Guo J, Zhang A. The renal and cardiovascular effects of natriuretic peptides. Adv Physiol Educ. 2017;41:179–185.

100. Yao X, Tian Z. Dyslipidemia and colorectal cancer risk: a meta-analysis of prospective studies. Cancer Causes Control. 2015;26:257–268.
101. Zamorano JL, Lancellotti P, Rodriguez Muñoz D, et al.; ESC Scientific Document Group. 2016 ESC Position Paper on cancer treatments and cardiovascular toxicity developed under the auspices of the ESC Committee for Practice Guidelines: The Task Force for cancer treatments and cardiovascular toxicity of the European Society of Cardiology (ESC). Eur Heart J. 2016;37:2768–2801. Erratum in: Eur Heart J. 2016. PMID: 27567406.
102. Zhang D, Ma QY, Hu HT, et al. β2-adrenergic antagonists suppress pancreatic cancer cell invasion by inhibiting CREB, NFκB and AP-1. Cancer Biol Ther. 2010;10:19–29.

7

Metabolomics in the Identification of New Biomarkers in Cardio-Oncology

Christian Cadeddu Dessalvi, Martino Deidda, Antonio Noto, Giuseppe Mercuro

Department of Medical Sciences and Public Health – University of Cagliari, Italy

****Correspondence to:***
Christian Cadeddu Dessalvi, MD
Department of Medical Sciences and Public Health,
University of Cagliari,
Cagliari, Italy
Email: cadedduc@unica.it

KEYWORDS: Biomarkers; Cardio-Oncology; Cardiotoxicity; Metabolomics; Precision Medicine.

7.1 Introduction

In the diagnosis of chemotherapy cardiotoxicity, biomarkers were immediately identified as the possible best tool that would allow the finest monitoring and prevention of this phenomenon. Cardiac biomarkers may be useful at different phases of cancer management, like estimating baseline risk prior to cancer treatment, during treatment in monitoring for early toxicity and to evaluate late side effects in survivors. However, compared to initial high expectations, most biomarkers showed inconclusive results as effective tools in the early diagnosis of cardiotoxicity (Zamorano et al. 2016). CV toxicity can be acute, subacute, or present itself years after chemotherapy or radiotherapy, involving several cardiac structures leading to heart failure (HF), valvular heart disease, coronary artery disease, arrhythmias, and pericardial disease. Biomarkers have been tested in different cancer settings showing

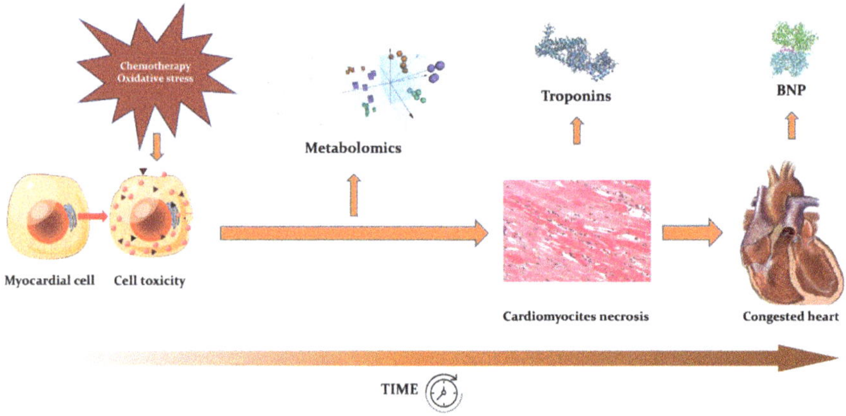

Figure 7.1 Timing of cardiotoxicity detection by conventional biomarkers and metabolomics.

encouraging results mostly in the prediction of left ventricular dysfunction and consequently in the development of HF (Cardinale et al. 2015).

Left ventricular impairment and HF is the most common cardiotoxicity deriving from cancer therapies. However, definition in trials and routine clinical practice vary and currently the role of cardiac biomarkers appears not clearly defined. Cardiac troponins (cTn) (both troponin I and troponin T) are the most studied biomarkers for the detection of early myocardial injury; however, a standardized timing of the blood samples and a specific cutoff for its clinical use has not been defined (Cardinale et al. 2000). To date, cardiotoxicity surveillance involving the detection of cardiac biomarkers appears to have a role only if combined with serial imaging, providing the most sensitive strategy to detect early toxicity and guide cardioprotective interventions (Fallah-Rad et al. 2011). Using this combined approach, cardiologists could support ongoing oncology treatment to completion in >85% of patients referred with cardiotoxicity from their current cancer treatment (Pareek et al. 2018).

More recently, the new omics sciences are emerging in this context with promising results.

Of all, metabolomics has provided the first encouraging results in animal models (Deidda et al. 2019) and the first data on small clinical trials show the ability of metabolomics to identify patients who are undergoing significant early cardiovascular toxicity (Cocco et al. 2020) (Figure 7.1).

7.2 Metabolomics

Metabolism, from the Greek μεταβολή or "change," is a term that indicates the set of life-sustaining chemical reactions by which the body converts nourishment into energy. It has a crucial role in balancing cellular physiology reactions (cellular growth and energy production) and cellular defense against metabolic aberrations (such as oxidative stress, toxicants, etc.). The biological mechanisms behind these processes are highly complex depending both on the genetic and epigenetic asset, and their final products are molecules named metabolites (Nielsen 2017).

Metabolites are low-molecular weight chemical with a mass range between 50 and 1500 daltons (Da), being either metabolic intermediates or metabolic end products, resulting from the cellular metabolism. Indeed, metabolites neither correlate "one-to-one" with genes nor originate from a single biochemical reaction; rather, multiple metabolites can originate from one enzyme (Zelezniak et al. 2014). In the human body, metabolites could result from human or microbial metabolism, environment, and diet. The identification of single metabolites in biological samples, such as glucose, creatinine, and many others, have been considered important for the diagnosis and monitoring of various diseases, such as diabetes mellitus, and kidney disease. However, for modern medicine, this approach is largely incomplete, hampering an overview of molecular mechanisms involved in a given disease. Metabolomics recognizes and measures metabolites detectable in a given biological sample, thus connecting genotype and phenotype by integrating the individual's epigenetic and genetic variation, environment, microbiota, and lifestyle each other (Holmes et al. 2008; Baker 2011).

The metabolomic approach consists of three sequential steps: (Zamorano et al. 2016) samples collection and storage, (Cardinale et al. 2015) samples analysis using an experimental technique, (Cardinale et al. 2000) and data analysis. Regarding the first step, the following rules should always be applied: collection of samples in a sterile container containing micromolar quantities of inorganic bacteriostatic agents such as sodium azide (0.01–0.1%) to avoid metabolic alterations due to bacterial metabolism; quick centrifugation of the samples at high speed to eliminate cellular debris, including active enzymes that can modify the metabolic content; early freezing of the samples at very low temperatures (−40 °C; −80 °C), up to the analysis; thawing the samples on ice to avoid rapid and harmful temperature variations (Gika et al 2008). Regarding the second step, many different technological platforms are used, such as high-performance liquid chromatography coupled with mass spectrometry (LC-MS), gas-chromatography coupled with mass spectrometry (GC-MS), and proton nuclear magnetic resonance spectroscopy (1H NMR). Based both on the technology performed and on the experimental design, a further classification concerns untargeted and targeted metabolomics. The former analysis includes all the detectable metabolites present in a biological sample, while the latter studies only specific class of metabolites, such as lipids, carbohydrates, organic acids, amino acids, and so on (Ribbenstedt et al. 2018). Finally, data analysis produces complex data matrices, consisting of the quantitative measurement of several hundred metabolites, which are subsequently analyzed with chemometrics methods, multivariate statistical tools that allow the extraction of biological, and physiological and clinical information. The conversion of metabolic patient-specific data into actionable clinical applications takes more than a robust platform technology. Despite the importance to interpret metabolomics data by the comparison between the individual metabotype with that of a reference population, clinical metabolomics testing should be based on an individualized approach (Hoffmann et al. 2011; Moyer et al. 2019).

Since biomarkers can be considered the key to the individualized treatment and precision medicine, metabolomics is basic for the discovery of novel biomarkers potentially useful in clinical practice and for deciphering alterations of the cellular functionality and metabolic pathway perturbations due to a given disease (Tolstikov et al. 2020). In fact, the individual metabolic profile, also called metabotype, is crucial to identify the susceptible individuals to cardiotoxicity, given the fact that the same pathological injury may originate different responses among individuals. An example is the diagnosis of heart failure (HF) due to cardiotoxicity (CTX), a leading cause of death worldwide. In a homogeneous group of such disease, usually treated by a conventional therapy, some of them may be found different only from the metabolomics point of view, suggesting that the disease and the response to the treatment are only identifiable at the molecular basis (Deidda et al. 2019). Metabolomics findings suggest that energy metabolism is a critical target in the development of this CTX form. Therefore, research aims to investigate these pathways for the identification of early markers of CTX and the development of innovative cardioprotective agents.

7.3 Metabolomics in Cardio-Oncology

The first study in this setting has been conducted by Andreadou et al. performing an NMR metabolomics profile of acute doxorubicin (DOX) induced CTX in a mouse model. After 3 days of DOX administration, a time sufficient to determine acute CTX at the cardiomyocyte level, the authors acquired 1H-NMR spectra of aqueous myocardial extracts. Authors found increased myocardial levels of acetate and succinate in DOX-treated samples, whereas branched-chain amino acids decreased, concluding that acetate and succinate could be useful as CTX biomarkers; moreover, oleuropein, a phenolic antioxidant derived from olive tree with documented cardioprotective effects, could reduce the distress of the energy metabolism (Andreadou et al. 2009).

The same researchers confirmed that oleuropein could have a role in the CTX prevention. In this study, metabolomic data were analyzed together with cardiac geometry and function evaluated by echocardiography, cardiac histopathology, nitroxidative stress, inflammatory cytokines, NO homeostasis (iNOS and eNOS expressions), and kinases involved in apoptosis (Akt and AMPK). The authors found a) reduced fractional shortening and left ventricular wall thickness in the DOX group, b) altered protein biosynthesis, c) an imbalance between the expression of iNOS and eNOS, and d) a perturbation in energy metabolism. However, Oleuropein seemed to prevent the DOX-mediated CTX inducing AMPK activation and iNOS suppression (Andreadou et al. 2009). It is noteworthy that, on the bases of identified metabolites, energy production pathways resulted involved in the CTX development in both the studies of Andreadu (Andreadou et al. 2009, 2014).

The centrality of these energetic pathways was confirmed by a GC–MS metabolomic study in which Tan *et al.* (2011) identified a fingerprint consisting of 24 metabolites involved in glycolysis, citrate cycle and metabolism of some amino acids and lipids in this rat model of DOX-induced CTX, evaluated as increases in creatine kinase (CK), CK-MB and lactate dehydrogenase (LDH) 3 days after DOX administration.

In order to reduce pirarubicin (THP) derived CTX, it has been developed a liposomal drug delivery system. Cong and colleagues studied the metabolic footprint of Sprague-Dawley rats' urine after three successive doses of liposome powder THP (L-THP) or free THP (F-THP). The metabonomic analysis confirmed that L-THP caused minimal metabolic changes compared to F-THP; moreover, subsequent doses of THP determined severe metabolic alterations, particularly at the level of energy production pathways. In detail, in the treatment groups was observed a significant downregulation of citrate (Krebs cycle), lactate (glycolysis), D-gluconate-1-phosphate (pentose phosphate), N-acetyl glutamine, and N-acetyl-DL-tryptophan (amino acid metabolism) (Cong et al. 2012).

A mouse model of CTX, analyzed by ultra-performance liquid chromatography–quadrupole time-of-flight mass spectrometry (UPLC–Q-TOF-MS), identified 39 biomarkers able to point out development of CTX, defined as severe heart damage evaluated by biochemical analysis and histopathological assessment. To filter out the biomarkers specific for CTX, the fingerprint was corrected for hepatotoxicity and nephrotoxicity, thus obtaining a panel consisting of 10 highly specific metabolites; among them, the most strongly specific resulted L-Carnitine, 19-hydroxydioxycortic acid, lysophosphatidylcholine (LPC) (14:0), and LPC (20:2). Furthermore, this panel showed to change before biochemical and histopathological alterations (Li et al. 2015).

To identify the early biomarkers of CTX induced by DOX treatment, a metabolomic study applying mass spectrometry and NMR spectroscopy was designed and carried out on male B6C3F1 mice, to whom 3 mg/kg DOX dose or saline were administered weekly for 2, 3, 4, 6, or 8 weeks; one week after the last dose, animals were sacrificed. At the myocardial level, an increase of 18 amino acids and 4 biogenic amines (acetylornithine, kynurenine, putrescine, and serotonin) was detected after a cumulative dose of 6 mg/kg; on the contrary, authors identified a myocardial lesion only at a cumulative dose of 18 mg/kg, and the cardiac pathology was highlighted at 24 mg/kg of cumulative dose. The metabolic analysis also revealed altered plasma levels of 16 amino acids, 2 biogenic amines (acetylornithine and hydroxyproline), and 16 acylcarnitines, whereas 5 acylcarnitines resulted in decreased cardiac tissue. It is important to highlight that plasma concentrations of lactate and succinate, two intermediates of the Krebs cycle, were significantly modified after a very low cumulative dose (6 mg/kg) (Schnackenberg et al. 2016), well before of the histological and clinical evidence.

To identify biomarkers of CTX, samples collected from rats after cyclophosphamide (CY) treatment were analyzed using UPLC–Q-TOF-MS. Metabolomic analysis showed altered levels of a dozen metabolites in the plasma of CY-treated group after 1, 3, and 5 days in comparison of the control group. Authors hypothesized that these molecules, involved in the metabolism of glycerol phospholipid and linoleic acid, may be implicated in the CTX induced by CY and suggested that it could determine increased myocardial oxygen consumption and impaired fatty acid β oxidation (Yin et al. 2015). In a study designed to evaluate the effects on human-induced pluripotent stem cell-derived cardiomyocytes (hiPSC-CMs) of DOX for 2 days or 6 days in a repeated way, ^{1}H-NMR spectroscopy was used to profile the culture medium. A single DOX exposure did not result in changes of the extracellular metabolites, whereas repeated exposures determined an impairment in the utilization of pyruvate and acetate, with an accumulation of formate. Moreover, during the washout from DOX were demonstrated a reversible effect and a restored utilization for pyruvate by hiPSC-CMs, while formate and acetate presented an irreversible effect. On these findings, the authors proposed a role for pyruvate, acetate, and formate as biomarkers of CTX induced by DOX (Chaudhari et al. 2017). In order to identify a biomarker of both DOX-induced CTX and cardioprotection by dexrazoxane (DZR), 96

BALB/c mice were randomly divided into two groups (tumor and control), each split into four treatment subgroups (control, DOX, DZR, and DOX plus DZR). A weight loss >20% established the moment to euthanize the animals. Metabolomic analysis showed a DOX administration fingerprint constituted by an increase in 5-hydroxylysine, 2-hydroxybutyrate, 2-oxoglutarate, and 3-hydroxybutyrate levels and a decrease in glucose, glutamate, cysteine, acetone, methionine, aspartate, isoleucine, and glycylproline levels. For its part, DZR treatment caused increased levels of lactate, 3-hydroxybutyrate, glutamate, and alanine and decreased levels of glucose, trimethylamine N-oxide, and carnosine. It is noteworthy that the authors suggested that their findings seem to confirm the importance of altered energy metabolism in the development of CTX (QuanJun et al. 2017).

Recently, tyrosine kinase inhibitors (TKI) have become an effective option for the treatment of a wide range of malignancies. CTX is a severe complication also of TKI use, probably due to their impact on specific cardiac metabolic pathways, with which this class of drugs interacts. To investigate the cardiotoxic effects of sorafenib, a non-targeted GC–MS metabolomics analysis was applied to the heart, skeletal muscle, liver, and plasma collected from FVB/N mice treated for two weeks with daily administration of sorafenib or vehicle control. Compared to controls, echocardiography of sorafenib-treated mice showed systolic dysfunction, while metabolomic analysis found significant changes in 11 metabolites, including a marked alteration in the metabolism of taurine/hypotaurine (Jensen et al. 2017a).

Another study aimed to evaluate the pathophysiology of CTX induced by TKIs was conducted on 10/group female FVB/N mice treated every day with sunitinib (40 mg/kg), erlotinib (50 mg/kg), or vehicle (control group) daily for a time of two weeks. The authors carried out a non-targeted GC–MS based metabolomic analysis on the specimens of heart, skeletal muscle, liver, and on serum. The sunitinib-treated mice showed, in comparison with the control group, a systolic dysfunction at echocardiography and a significant decrease in levels of docosahexaenoic acid, arachidonic acid/eicosapentaenoic acid, O-phosphocolamine, and 6-hydroxynicotinic acid at the metabolomic analysis. On the other hand, erlotinib did not determine a decrease in systolic function but only an increase in the levels of spermidine. The authors highlighted that their study suggests a link between CTX due to sunitinib treatment, polyunsaturated FA depletion, and inflammatory mediators (Jensen et al. 2017b).

Aimed to investigate the metabolic changes induced in cardiac cells by the exposure of the heart to ionizing radiation, Gramatyka and colleagues studied irradiated human cardiomyocytes using high-resolution magic-angle-spinning nuclear magnetic resonance techniques (HR-MAS NMR). The metabolomic analysis found changes in lipids, threonine, glycine, glycerophosphocholine, choline, valine, iso-leucine and glutamate, as well as impaired metabolism of glutathione and taurine. Authors concluded that ionizing radiations are able to alter the c cardiomyocytes metabolic pathways even at low doses, which potentially do not affect cell viability (Gramatyka et al. 2018).

A more recent study by Unger and colleagues carried out a metabolomic and lipidomic analysis on male Sprague Dawley (SD) rats sham irradiated or subjected to receive fractionated doses (9 Gy per day × 5 days) of targeted X-ray heart radiation; plasma and left ventricle heart tissue samples were collected and analyzed. Moreover, authors applied high-resolution GC_MS to profile the plasma samples of esophageal cancer patients treated with high dose thoracic RT. Study results showed commonalities between the metabolic alterations induced by radiations in the rat model and in cancer patients, including steroid hormone biosynthesis and vitamin E metabolism. Moreover, these findings were used to develop algorithms able to classify the risk of developing radiation-induced heart disease in patients (Unger et al. 2020).

7.4 Metabolomics Profiling Cardioprotective Strategies

In clinical practice, traditional biomarkers have been used to monitor the efficacy of several cardioprotective strategies (Gulati et al. 2017; Avila et al. 2018; Cardinale et al. 2018). The International Cardio Oncology Society-One (ICOS-1) trial investigated an ACE-inhibitor as enalapril used in patients receiving anthracyclines either as primary prevention before anthracyclines or after a detection of a troponin rise in a biomarker-guided cardioprotective strategy (Cardinale et al. 2018). The study established that low dose enalapril (mean 5 mg/day) did not affect anthracycline-related increase in troponin, but it did reduce LV remodeling in both study arms.

Another primary prevention trial with candesartan (PRADA trial) confirmed in a breast cancer cohort receiving anthracycline the inability to prevent troponin release although reducing adverse remodeling (Gulati et al. 2017). On the contrary, primary prevention with beta blockers administered before anthracycline treatment showed to be able to reduce the troponin rise although not affecting significantly LV remodeling (Gulati et al. 2017; Avila et al. 2018).

However, since several studies have shown that in this setting the findings are inconstant based on the type of biomarker and on whether biomarkers are used pre-, during, or post-chemotherapy, more data is needed to reach significant conclusions (Sandri et al. 2005; Cardinale et al. 2017). Moreover, several studies including ICOS-1 (Cardinale et al. 2018) have demonstrated that troponin may not always be the most suitable biomarker by showing that it is possible for it to rise independently of cardiotoxicity.

Metabolomic could be of help also in this setting with promising results to better understand cardioprotection physiopathology, although, so far, only experimental models have been tested. The potential cardioprotective effect of losartan against sorafenib-induced cardiotoxicity has been studied with a GC-MS non-targeted based metabolite profiling of rat plasma conducted together with echocardiographic exam. Sorafenib induced significant alterations in indexes of myocardial contractility and relaxation, reversed by the co-administration of losartan. Sorafenib-induced CTX was characterized by elevation in some metabolites levels, including urea and fatty acids; however, only glycine and lactic acid resulted statistically significant. It is noteworthy that losartan demonstrated to be able to restore these changes, resulting in a significantly reduced glycine, urea, and some fatty acids levels (cis-vaccenic acid, oleic acid, stearic acid, and undecanoic acid). Authors concluded that losartan could be effective in determining a protective effect against CTX induced by sorafenib (Abdelgalil et al. 2020).

Yoon and colleagues performed an ^{1}H-NMR-based metabolomic analysis to investigate the cardioprotective effects of spinochrome D (SpD) evaluating its effects on human cardiomyocytes and human breast cancer cells against 24/48 h exposure to 0.1 μM DOX. SpD did not determine any effect on DOX anticancer properties but seemed to protect AC16 cells from its toxicity. Furthermore, SpD was able to determine different mitochondrial membrane potentials and calcium localization between cardiomyocytes and cancer cell lines, suggesting a possible explanation for the observed SpD role against CTX due to DOX administration. A decrease in acetate, glutamine, serine, uracil, glycerol levels, and an increase in those of glutamate, isoleucine, O-phosphocholine, taurine, myo-inositol, glutathione, and sn-glycero-3-phosphocholine were found; moreover, glutathione metabolism resulted the most significantly altered pathway by SpD (Yoon et al. 2018).

7.5 Conclusion

Metabolomics proved to be an effective tool for the early diagnosis of chemotherapy-related CTX being able to identify the first signs of metabolic pathways alteration (Figure 7.1). Most of the studies seem to point out the changes in energy metabolism as the most affected; however, several other findings show the ability of this technique in establishing a specific metabolic predisposition for antiblastic drug induced CTX.

The basic and translational metabolomic approach, recognizing specific metabolic profiles related to the risk of CTX, will make *a priori* stratification and very early identification of the CTX risk possible, well before the onset of significant alterations reported by commonly used biomarkers, which are mostly indexes of occurred cardiac damage. Indeed, studies on animal models and in small cohort of patients seem ready to translate this initial perception of efficacy into clinical evidence. Moreover, great expectations rely on the large number of pathophysiological data provided by metabolomics as well as by other unconventional strategies (Madonna et al. 2015; Tocchetti et al. 2019), to be able to identify a highly effective and individualized therapeutic strategy, able to prevent and treat CTX (Cadeddu et al. 2016).

References

1. Abdelgalil AA, Mohamed OY, Ahamad SR, et al. The protective effect of losartan against sorafenib induced cardiotoxicity: Ex-vivo isolated heart and metabolites profiling studies in rat. Eur J Pharmacol. 2020;882:173229.
2. Andreadou I, Mikros E, Ioannidis K, et al. Oleuropein prevents doxorubicin-induced cardiomyopathy interfering with signaling molecules and cardiomyocyte metabolism. J Mol Cell Cardiol. 2014;69:4–16.
3. Andreadou I, Papaefthimiou M, Zira A, et al. Metabonomic identification of novel biomarkers in doxorubicin cardiotoxicity and protective effect of the natural antioxidant oleuropein. NMR Biomed. 2009;22:585–92.
4. Avila MS, Ayub-Ferreira SM, de Barros Wanderley MR Jr, et al. Carvedilol for Prevention of Chemotherapy-Related Cardiotoxicity: The CECCY Trial. J Am Coll Cardiol. 2018;71.:2281–2290.
5. Baker M. Metabolomics: from small molecules to big ideas. Nat Methods 2011; 2:219–221.
6. Cadeddu C, Mercurio V, Spallarossa P, et al. Preventing antiblastic drug-related cardiomyopathy: old and new therapeutic strategies. J Cardiovasc Med (Hagerstown). 2016;17S1:e64-e75.
7. Cardinale D, Biasillo G, Salvatici M, et al. Using biomarkers to predict and to prevent cardiotoxicity of cancer therapy. Expert Rev Mol Diagn. 2017;17:245–256.
8. Cardinale D, Ciceri F, Latini R, et al.; ICOS-ONE Study Investigators. Anthracycline-induced cardiotoxicity: A multicenter randomised trial comparing two strategies for guiding prevention with enalapril: The International CardioOncology Society-one trial. Eur J Cancer. 2018;94:126–137.

9. Cardinale D, Colombo A, Bacchiani G, et al.. Early detection of anthracycline cardiotoxicity and improvement with heart failure therapy. Circulation. 2015;131:1981–8.
10. Cardinale D, Sandri MT, Martinoni A, et al. Left ventricular dysfunction predicted by early troponin I release after high-dose chemotherapy. J Am Coll Cardiol. 2000;36:517–22.
11. Chaudhari U, Ellis JK, Wagh V, et al. Metabolite signatures of doxorubicin induced toxicity in human induced pluripotent stem cell-derived cardiomyocytes. Amino Acids. 2017;49:1955–1963.
12. Cocco D, Ferro EG, Ricci S, et al. Defining the metabolomic profile associated with early cardiotoxicity in patients with breast cancer treated with anthracyclines Eur Heart J 2020;41:S2,ehaa946.3289.
13. Cong W, Liang Q, Li L, et al. Metabonomic study on the cumulative cardiotoxicity of a pirarubicin liposome powder. Talanta. 2012;89:91–8.
14. Deidda M, Mercurio V, Cuomo A, et al. Metabolomic Perspectives in Antiblastic Cardiotoxicity and Cardioprotection. Int J Mol Sci. 2019;20:4928.
15. Fallah-Rad N, Walker JR, Wassef A, et al. The utility of cardiac biomarkers, tissue velocity and strain imaging, and cardiac magnetic resonance imaging in predicting early left ventricular dysfunction in patients with human epidermal growth factor receptor II-positive breast cancer treated with adjuvant trastuzumab therapy. J Am Coll Cardiol. 2011;57:2263–70.
16. Gika HG, Theodoridis GA, Wilson ID. Liquid chromatography and ultra-performance liquid chromatography-mass spectrometry fingerprinting of human urine: sample stability under different handling and storage conditions for metabonomics studies. J Chromatogr A. 2008;1189:314–22.
17. Gramatyka M, Skorupa A, Sokół M. Nuclear magnetic resonance spectroscopy reveals metabolic changes in living cardiomyocytes after low doses of ionizing radiation. Acta Biochim Pol. 2018;65:309–318.
18. Gulati G, Heck SL, Røsjø H, et al. Neurohormonal Blockade and Circulating Cardiovascular Biomarkers During Anthracycline Therapy in Breast Cancer Patients: Results From the PRADA (Prevention of Cardiac Dysfunction During Adjuvant Breast Cancer Therapy) Study. J Am Heart Assoc. 2017;6:e006513.
19. Hoffmann W, Krafczyk-Korth J, Völzke H, et al. Towards a unified concept of individualized medicine. Per Med. 2011;8:111–13.
20. Holmes E, Wilson ID, Nicholson JK. Metabolic phenotyping in health and disease. Cell 2008;134:714–7.

21. Jensen BC, Parry TL, Huang W, et al. Effects of the kinase inhibitor sorafenib on heart, muscle, liver and plasma metabolism in vivo using non-targeted metabolomics analysis. Br J Pharmacol. 2017a;174:4797–4811.
22. Jensen BC, Parry TL, Huang W, et al. Non-Targeted Metabolomics Analysis of the Effects of Tyrosine Kinase Inhibitors Sunitinib and Erlotinib on Heart, Muscle, Liver and Serum Metabolism In Vivo. Metabolites. 2017b;7:31.
23. Li Y, Ju L, Hou Z, et al. Screening, verification, and optimization of biomarkers for early prediction of cardiotoxicity based on metabolomics. J Proteome Res. 2015;14:2437–45.
24. Madonna R, Cadeddu C, Deidda M, et al. Cardioprotection by gene therapy: A review paper on behalf of the Working Group on Drug Cardiotoxicity and Cardioprotection of the Italian Society of Cardiology. Int J Cardiol. 2015;191:203–10.
25. Moyer AM, Matey ET, Miller VM. Individualized medicine: Sex, hormones, genetics, and adverse drug reactions. Pharmacol Res Perspect. 2019; 7:e00541.
26. Nielsen J. Systems Biology of Metabolism: A Driver for Developing Personalized and Precision Medicine. Cell Metab. 2017;25:572–579.
27. Pareek N, Cevallos J, Moliner P, et al. Activity and outcomes of a cardio-oncology service in the United Kingdom-a five-year experience. Eur J Heart Fail. 2018;20:1721–1731.
28. QuanJun Y, GenJin Y, LiLi W, et al. Protective Effects of Dexrazoxane against Doxorubicin-Induced Cardiotoxicity: A Metabolomic Study. PLoS One. 2017;12:e0169567.
29. Ribbenstedt A, Ziarrusta H, Benskin JP. Development, characterization and comparisons of targeted and non-targeted metabolomics methods. PLoS One. 2018;13:e0207082.
30. Sandri MT, Salvatici M, Cardinale D, et al. N-terminal pro-B-type natriuretic peptide after high-dose chemotherapy: a marker predictive of cardiac dysfunction? Clin Chem. 2005;51:1405–10.
31. Schnackenberg LK, Pence L, Vijay V, et al. Early metabolomics changes in heart and plasma during chronic doxorubicin treatment in B6C3F1 mice. J Appl Toxicol. 2016;36:1486–95.
32. Tan G, Lou Z, Liao W, et al. Potential biomarkers in mouse myocardium of doxorubicin-induced cardiomyopathy: a metabonomic method and its application. PLoS One. 2011;6:e27683.
33. Tocchetti CG, Cadeddu C, Di Lisi D, et al. From Molecular Mechanisms to Clinical Management of Antineoplastic Drug-Induced Cardiovascular Toxicity: A Translational Overview. Antioxid Redox Signal. 2019;30:2110–2153.

34. Tolstikov V, Moser AJ, Sarangarajan R, et al. Current Status of Metabolomic Biomarker Discovery: Impact of Study Design and Demographic Characteristics. Metabolites. 2020;10:224.
35. Unger K, Li Y, Yeh C, et al. Plasma metabolite biomarkers predictive of radiation induced cardiotoxicity. Radiother Oncol. 2020;152:133–145.
36. Yin J, Xie J, Guo X, et al. Plasma metabolic profiling analysis of cyclophosphamide-induced cardiotoxicity using metabolomics coupled with UPLC/Q-TOF-MS and ROC curve. J Chromatogr B Analyt Technol Biomed Life Sci. 2016;1033–1034:428–435.
37. Yoon CS, Kim HK, Mishchenko NP, et al. Spinochrome D Attenuates Doxorubicin-Induced Cardiomyocyte Death via Improving Glutathione Metabolism and Attenuating Oxidative Stress. Mar Drugs. 2018;17:2.
38. Zamorano JL, Lancellotti P, Rodriguez Muñoz D, et al.; ESC Scientific Document Group. 2016 ESC Position Paper on cancer treatments and cardiovascular toxicity developed under the auspices of the ESC Committee for Practice Guidelines: The Task Force for cancer treatments and cardiovascular toxicity of the European Society of Cardiology (ESC). Eur Heart J. 2016;37:2768–2801. Erratum in: Eur Heart J. 2016: PMID: 27567406.
39. Zelezniak A, Sheridan S, Patil KR. Contribution of Network Connectivity in Determining the Relationship between Gene Expression and Metabolite Concentration Changes. PLoS Comput Biol. 2014;10:e1003572.

8

Imaging in Cardio-Oncology

Roberta Manganaro* **and Concetta Zito**

Department of Clinical and Experimental Medicine
Unit of Cardiology – University of Messina, Messina Italy

****Correspondence to:***
Roberta Manganaro, MD
Department of Clinical and Experimental Medicine
Unit of Cardiology - University of Messina, Messina Italy
Tel: +39-090-221-2969
E-mail: manganaro.roberta@gmail.com

KEYWORDS: Cardiotoxicity; Multimodality Imaging; Echocardiography; Myocardial Strain; Risk Stratification; Follow Up.

8.1 Introduction

Cardiovascular (CV) complications in cancer patients present a growing medical problem, causing substantial premature mortality in this population. Myocardial dysfunction and heart failure (HF), frequently described as cardiotoxicity, are the most concerning cardiovascular complications of cancer therapies and cause an increase in morbidity and mortality. Contemporary cardiac imaging plays a pivotal role for baseline risk stratification, timely diagnosis of early CV disease and of cardiac dysfunction, both during and following treatment, for the identification of cancer patients who may benefit from cardioprotective treatments whilst continuing oncology treatment, and prognostication to select cancer patients who may require long-term CV follow-up (Čelutkienė et al. 2020).

Echocardiography is a crucial tool in these settings: for risk stratification, at baseline and during follow-up of cancer patients to early identify left ventricular (LV) impairment and closely monitoring cardiac function throughout the oncological treatment. Different definitions of cancer therapy-related cardiac dysfunction (CTRCD) based on echocardiographic LV ejection fraction (EF) assessment are proposed in guidelines, position statement, and oncology trials (Čelutkienė et al. 2020; Zamorano et al. 2016; Plana et al. 2014). According to the 2016 ESC cardio-oncology position statement, CTRD is defined as a decrease in the LVEF of >10 percentage points to a value <50%. (2) Current echocardiography recommendations set low normal value of two-dimensional (2D) LVEF as 54% for women and 52% for men (Lang et al. 2015); (4) thus, in the previous 2014 ASE/EACVI Expert Consensus, a reduction of LVEF below 53% was classified as abnormal (Plana et al. 2014). (3) However, a decrease of LVEF occurs only after a relevant ultrastructural injury, and, thus, it does not allow the detection of early cardiotoxicity (CTX) (Ewer and Lenihan 2008; Sawaya et al. 2012). For this purpose, new parameters, such as global longitudinal strain (GLS), were developed in the last few years, considerably increasing the reliability and sensitivity of echocardiography.

Furthermore, other imaging modalities, such as cardiac magnetic resonance (CMR), computed tomography (CT) and nuclear testing may offer additional information, especially in case of suboptimal quality of echocardiographic images or when tissue characterization is needed, or in the search of cancer therapy-related coronary artery disease (CAD).

8.2 Conventional Echocardiography

8.2.1 LV systolic and diastolic function

Echocardiography is the main imaging modality in cardio-oncology, due to the known availability, feasibility, low cost, easy repeatability, and safety of the method. It allows measurement of LV and right ventricle (RV) dimensions, systolic and diastolic function, and also a comprehensive evaluation of cardiac valves, aorta, and pericardium. It is of paramount importance to perform echocardiography before starting potentially cardiotoxic therapy in every cancer patient as a baseline for monitoring and for risk stratification. Changes in LVEF should be better evaluated in these patients by comparing baseline and follow-up studies. The most relevant parameters for a focused echocardiographic assessment of cardiac function are shown in Table 8.1.

The most commonly used parameter for monitoring LV function with echocardiography is LVEF. Calculation of LVEF should be done with the best method available in every echocardiography laboratory. Efforts should be done to adopt the same method to determine LVEF whenever possible during treatment and surveillance after treatment. Unless three-dimensional

Table 8.1 Echocardiographic parameters relevant for cardio-oncology surveillance.

Parameters	Clinical significant changes
LV size and function	
• LVEF by Simpson's 2D, or (semi) automatic 3D	Drop >10% (percentage points) for 2D, >5% for 3D from pre-treatment value
• 2D/3D GLS, GCS	Relative reduction by >10%–15% from pre-treatment value and to below lower limit of normal
• LV 2D/3D systolic and diastolic volumes	Increase by 15 mL for ESV, 30–35 mL for EDV
RV function, pulmonary artery pressure, and volemia	
• Markers of systolic RV function	TAPSE <1.7 cm, FAC <35%, RV free wall strain <20%, 3D RVEF <45%
• Velocity of TR	Peak systolic TR velocity > 2.8 m/s
• IVC diameter, collapse on inspiration	Dilatation >2.1 cm or narrowing <1.3 cm

2D, two-dimensional; 3D, three-dimensional; EDV, end-diastolic volume; ESV, end-systolic volume; FAC, fractional area change; GCS, global circumferential strain; GLS, global longitudinal strain; IVC, inferior vena cava; LV, left ventricular; LVEF, left ventricular ejection fraction; RV, right ventricular; RVEF, right ventricular ejection fraction; TAPSE, tricuspid annular plane systolic excursion; TR, tricuspid regurgitation. Modified from Čelutkienė *et al.* (2020).

echocardiography (3DE) is used, which is the best echocardiographic method for measuring LVEF when endocardial definition is clear, 2D biplane Simpson method is recommended for estimation of LV volumes and EF in these patients.

When transthoracic echocardiographic image quality is inadequate for the application of Simpson's method, which is more common in cancer patients who have previously undergone left breast or left chest surgery and/or radiotherapy, and sometimes in very cachectic patients, adding contrast media can be considered for serial monitoring of LV size and function.

The role of LVEF was extensively questioned because of its low sensitivity for the detection of small changes in left ventricle contractility (Dodos et al. 2008; Tassan-Mangina et al. 2006; Di Lisi et al. 2011), particularly if the changes are limited to few segments. Therefore, a careful analysis of regional alterations is strongly recommended beyond the LVEF assessment that should be associated with assessment of the wall motion score index.

As previously reported, CTRCD is defined as a decrease in the LVEF of > 10 percentage points, to a value below the lower limit of normal (Zamorano et al. 2016; Plana et al. 2014). This decrease should be confirmed by repeated cardiac imaging, performed 2–3 weeks after the baseline diagnostic study showing the initial decrease in LVEF. LVEF decrease may be further categorized as symptomatic or asymptomatic or with regard to reversibility (Plana et al. 2014).

Moreover, concurrent measurement of blood pressure may help to avoid misinterpretations in cases of blood pressure and blood volume changes due to fluid excess during intravenous chemotherapy or fluid loss due to adverse reactions.

Diastolic parameters have not been found to be prognostic of CTRCD. In particular, the use of the E/e′ ratio remains questionable in the oncological setting, as E and e′ velocity fluctuations in these patients could be the consequence of changes in loading conditions as a result of side effects associated with the chemotherapy (nausea, vomiting, and diarrhea) more than the result of a real change in LV diastolic performance. However, the ASE/EACVI Expert Consensus suggests that a conventional assessment of left ventricle diastolic function, including the grading of diastolic function and non-invasive estimation of LV filling pressures, should be added to assessments of LV systolic function (Plana et al. 2014).

8.2.2 RV function

Data on RV remodeling and dysfunction in oncology patients remain scarce and the prognostic role value of RV dysfunction has not been demonstrated in patients undergoing chemotherapy; however, RV evaluation should be part of echocardiographic investigation in oncological patients. In fact, RV may be damaged in cancer patients for neoplastic involvement (primary or metastatic) or as a result of the cardiotoxic effects of chemotherapy.

There are particular cardiotoxic cancer treatments that may specifically cause pulmonary arterial hypertension (dasatinib) (Montani et al. 2012) and/or RV dysfunction [anthracyclines (AC), trastuzumab, cyclophosphamide, and dasatinib]. A significant reduction of RV longitudinal strain has been shown within 3 months of AC therapy and after 6 months of trastuzumab use in HER2-positive breast cancer patients.

RV function and pulmonary artery pressure should be assessed at pre-treatment baseline and subsequently during echocardiographic surveillance. The frequency of monitoring depends upon the severity of the pre-existing pulmonary arterial hypertension or RV dysfunction and the risk of cardiotoxicity analogously to the monitoring of LV systolic dysfunction.

Echocardiographic evaluations of the right chambers in patients receiving cardiotoxic therapeutics should include the following measurements: basal RV diameter and right atrium area, tricuspid annular plane systolic excursion, peak of tricuspid annulus systolic velocity by tissue Doppler imaging (TDI), and fractional area change (Plana et al. 2014). When available, RV free wall strain and RV EF by 3DE should also be performed (see Table 8.1; Čelutkienė et al. 2020).

8.2.3 Valvular heart disease

Although chemotherapeutic agents seem not to affect cardiac valves, valvular heart disease can be observed in oncological patients. They can be pre-existing valve lesions or valve diseases secondary to concomitant or previous radiation therapy (RT), to severe infection as a complication of chemotherapy, or to CTRCD (Plana et al. 2014; Lancilotti et al. 2013). Long-term follow-up studies have shown progressive valvular heart disease in some patients, especially those with mediastinal radiation that includes the heart. In addition, these studies have observed that chemotherapy, particularly with anthracyclines, seemed to increase the risk valve damage (Crawford 2016). A Norwegian study that started in 1993 and was updated in 2009 studied 116 Hodgkin's lymphoma survivors at a mean of 10 years after radiation therapy and in some anthracycline therapy, and showed that 31% had moderate valve regurgitation (Lund et al. 1996). The subsequent follow-up of 51 of these patients at a mean follow-up of 22 years showed progressive valvular heart disease; probably, in these patients, cardiac anthracycline toxicity led to ventricular dysfunction and secondary or functional valve regurgitation.

Radiation therapy alone (27%) increased the risk of valve disease (HR: 6.6) in the group from the Netherlands reporting on the cardiovascular disease risk of Hodgkin's lymphoma patients after 40 years of follow-up (median 20 years) in 2524 patients treated with radiation and AC containing chemotherapy. In those treated with AC-based chemotherapy alone (6.7%), there was a significantly increased risk of valve disease (HR: 1.5) (van Nimwegen et al. 2015).

Furthermore, Murbraech *et al.* assessed the prevalence and associated risk factors for valvular heart disease in a cross-sectional national study of all adult lymphoma survivors after high-dose chemotherapy containing an anthracycline and autologous stem cell transplantation from 1987 to 2008. They found that age >50 years, women, > 3 lines of chemotherapy, and cardiac radiation were independent predictors of valve disease. In those treated by chemotherapy, only age >50 years and >3 lines of chemotherapy were associated significantly with valve disease. The authors concluded that AC-containing chemotherapy alone is associated with valvular heart disease due to valve degeneration. It can be postulated that chemotherapy does have direct cellular toxicity in mature cells that do not divide frequently such as the myocardium. Perhaps, valvular endothelium can be damaged as well, leading to scarring, leaflet retraction, and thickening, which could cause regurgitation and, eventually, stenosis. Further research need to be done to elucidate this intriguing concept. At this time, it seems prudent to consider valve disease in long-term lymphoma survivors whether they received radiation or chemotherapy or especially both at high doses as a young person (Murbraech et al. 2016).

Moreover, in patients with advanced malignant tumors, non-bacterial thrombotic or marantic endocarditis may occur, particularly in left sided valves (Edoute et al. 1997; Eiken et al. 2001).

Echocardiography remains the first-line and gold standard imaging modality for the assessment of valvular heart disease, allowing qualitative and quantitative evaluation of both stenotic and regurgitant valves. Patients with significant baseline or changing valvular findings during chemotherapy require more frequent serial echocardiographic examinations.

8.2.4 Pericardial disease

Pericardial involvement is quite frequent in oncological patients. It may be secondary to cardiac metastasis or may be a consequence of radiotherapy and/or chemotherapy (Gaya and Ashford 2005). (18, 19) Several chemotherapy agents were related to pericardial disease: anthracyclines (Casey et al. 2012), cyclophosphamide (Katayama et al. 2009), methotrexate (Savoia et al. 2010), arsenic trioxide (Huang et al. 1998), and, less frequently, 5-fluorouracil (Çalık et al. 2012), docetaxel (Dogan et al. 2017), and tyrosine-kinase inhibitors (Agrawal et al. 2015).

Pericardial disease induced by chemotherapy generally manifests as pericarditis, with or without associated myocarditis. Pericarditis can arise acutely during radiotherapy, leading to later pericardial constriction which typically presents over 10 years following treatment.

Echocardiography is the first line imaging modality for assessment of pericardial involvement. The echocardiographic findings in these patients may be entirely normal or show clear evidence of a pericardial effusion. The pericardial effusion should be quantified and graded using recognized methods to allow comparisons in subsequent evaluations. Echocardiographic and Doppler signs of cardiac tamponade should be investigated, particularly in patients with malignant effusions. When constrictive pericarditis is suspected, echocardiographic signs of constriction should be explored according to published guidelines (Klein et al. 2013; Cosyns et al. 2015). Other imaging modalities, such as computed tomography or CMR, may be a useful diagnostic complement. CMR is particularly helpful in determining the presence of late gadolinium enhancement for the identification of pericardial inflammation.

8.3 Advanced Echocardiography

8.3.1 Myocardial deformation imaging

The definition of CTRCD relies on the estimation of EF; however, the known limits of this parameter, especially if obtained by 2D echo, could lead to underestimation of early cardiotoxicity. The normal heart has tremendous recruitable contractile ability, although this is not adequately appreciated (Ewer and Lenihan 2008). Thus, a decrease of EF is a marker of advanced damage, occurring when the heart is no longer able to compensate. Moreover, an interesting study by Thavendiranathan *et al.* (2014) showed that the interoperator variability of EF was about 10% in the assessment of LV systolic function in patients treated by chemotherapy.

Myocardial deformation imaging allows to overcome the uncertain sensibility of LVEF in the evaluation of early impairment of systolic function in patients undergoing anticancer therapy (Sawaya et al. 2011; Jurcut et al. 2008). Myocardial strain can be studied using different ultrasound techniques, including Doppler strain imaging (DSI) and 2D and 3D speckle-tracking echocardiography (STE) (Mor-Avi et al. 2011). DSI was the first method used, and it was more sensitive than LVEF assessment in recognizing LV systolic dysfunction caused by chemotherapy and radiotherapy in adults and children (Hare et al. 2009; Erven et al. 2013). However, this tool exhibited significant limitations, such as low reproducibility, angle dependency, limited spatial resolution, a high sensitivity to signal noise, and high interobserver variability. STE was developed to overcome these limitations. Global longitudinal strain (GLS) has emerged as a new marker of subclinical ventricular dysfunction demonstrating stronger association with prognosis than LVEF (Zito et al. 2018, 2016). It starts to be impaired at an early stage of myocardial damage. However, circumferential strain compensates the reduction in contractile function, resulting in the preservation of LVEF until a later stage (Kraigher-Krainer et al. 2014).

Several studies provided information on serial evaluations of cardiac function before and after oncologic treatments by comparing GLS with LVEF (Stoodley et al. 2011; Kang et al. 2013). They found that GLS was the most sensitive and specific measurement for the detection of subclinical myocardial injury early after anthracycline exposure. GLS decreased significantly without any reduction in LVEF. Other studies showed that GLS reduction in patients treated with anthracyclines anticipates changes in LVEF, providing fundamental information for an early risk stratification of these subjects (Poterucha et al. 2012; Charbonnel et al. 2017).

Negishi *et al.* (2013) found that the strongest predictor of CTRCD was delta 2D-based GLS measured at the six-month visit, with a cutoff value of 11% point decrease (95% CI 8.3, 14.6). On this basis, the Expert Consensus from the American Society of Echocardiography and the European Association of Cardiovascular Imaging incorporate GLS in the algorithm for the early detection of subclinical LV dysfunction (Plana et al. 2014). According to this document, measurements of GLS during chemotherapy should ideally be compared with baseline value, and a relative percentage reduction of GLS of less than 8% from baseline is not meaningful, but more than 15% from baseline is very likely abnormal. Furthermore, it is strongly recommended to use the same vendor-specific ultrasound machine for longitudinal follow-up of patients with cancer.

The regionality of myocardial cardiotoxicity was investigated in 19 children at the midpoint and at the end of their anthracycline treatment. The authors found a septal and apical pattern, which was partially improved at the end of the treatment (Poterucha et al. 2012). These results were recently confirmed in a multicentric study on breast cancer patients underwent anthracyclines-based chemotherapy (Zito et al. 2021).

A recent study in 116 patients with human epidermal growth factor receptor 2 (HER2)-positive breast cancer supported the serial surveillance using GLS to guide cardioprotection and maintain patients on uninterrupted trastuzumab therapy (Santoro et al. 2019).

A definite recommendation on strain-based cardioprotection strategy has not been yet established, even if encouraging results in this field have been obtained (Santoro et al. 2019). The ongoing multicenter, randomized, controlled trail SUCCOUR (Strain Surveillance During Chemotherapy for Improving Cardiovascular Outcome) will provide more insights on the impact of strain-guided cardioprotective therapy in patients at risk of cardiotoxicity during cancer therapy (Negishi et al. 2018).

Therefore, GLS surveillance may become a more sensitive strategy for early detection of cardiotoxicity and guide timing of cardioprotective treatment.

Finally, other cancer drugs may cause different forms of myocardial toxicity where LVEF reduction is not the primary manifestation. For example, immune check-points (ICIs) cause myocarditis, which can lead to severe HF, cardiogenic shock, and death, but in 38% of cases, they may also occur even without a fall in LVEF. Thus, decision-making concerning the continuation or interruption of such potentially life-saving therapy should no longer rely solely on the single, surrogate echocardiographic parameter (LVEF) which mainly reflects changes in LV volumes, rather than function. GLS analysis in these patients seems to be very promising with a strong correlation with cardiac edema/fibrosis identified by MRI (Čelutkienė et al. 2020).

8.4 Three-Dimensional Echocardiography (3DE)

3DE is likely to become more widely accepted in routine practice due to improved image acquisition and the implementation of semiautomated or fully automated analysis algorithms. The feasibility of 3D LVEF in breast cancer patients with adequate echocardiographic images was 88% at baseline and 66% after AC therapy, reduced during follow-up due to concomitant radiotherapy (RT), left mastectomy, left breast prosthesis, and other patient factors (Narayan et al. 2017).

3DE is more accurate than 2DE in LV volumes and EF measurements, showing a precision that is comparable to CMR. (47) The advantages include better accuracy and reproducibility and lower temporal resolution compared to 2DE. 3DE showed improved accuracy over 2DE in detecting CMR-derived EF < 55% in survivors of childhood cancer (Armstrong et al. 2012; Toro-Salazar et al. 2016). These data may be explained by the fact that 3DE volume measurements are not based on geometric assumptions of LV shape and are not affected by apical foreshortening. Moreover, the automated or semiautomated method for the identification of the LV endocardium, compared with the manual tracing of endocardial contour that is required by the 2D method, provides a more accurate estimation of LV volumes (Jenkins et al. 2009; Muraru et al. 2010). Thus, LV systolic function should be analyzed by 3DE for monitoring of cardiac toxicity when available. However, its dependency on good image quality, time-consuming processing, and the need for operator training limit its widespread use.

8.5 Assessment of Cardiotoxicity Risk and Echocardiographic Surveillance According to Anticancer Treatment

The recent European Position Paper on cardiovascular imaging in cancer patients recommends a personalized approach according to patient's baseline risk of cardiotoxicity in order to balance rational use of resource

Table 8.2 Assessment of cardiotoxicity risk.

Therapy-related factors	Patient-related factors
Low risk	
• Lower dose AC (e.g., doxorubicin <200 mg/m^2, epirubicin <300 mg/m^2), liposomal formulations. • Trastuzumab without AC	• Age >18 and <50 years
Medium risk	
• Modest-dose AC (doxorubicin 200–400 mg/m^2 and epirubicin 300–600 mg/m^2) • AC followed by trastuzumab • VEGF tyrosine kinase inhibitors • Second- and third-generation Bcr-Abl tyrosine kinase inhibitors • Proteasome inhibitors • Combination immune checkpoint inhibitors	• Age 50–64 years • 1–2 CV risk factors such as hypertension, dyslipidemia, obesity, insulin resistance, smoking, etc.
High risk	•
• Simultaneous AC and trastuzumab • High-dose AC (doxorubicin ≥400 mg/m^2 or epirubicin ≥600 mg/m^2) • Modest-dose AC plus left chest radiation therapy • Elevated cardiac troponin post-AC prior to HER2-targeted therapy • High-dose radiation therapy to central chest including heart in radiation field ≥30 Gy • VEGF tyrosine kinase inhibitors following previous AC chemotherapy	• Age ≥65 years • >2 CV risk factors as hypertension, dyslipidemia, obesity, smoking, diabetes, etc. • Underlying CV disease: CAD, PAD, CMP, severe VHD, heart failure • Reduced or low-normal LVEF (50%–54%) pre-treatment • Prior cancer therapy

Abr, active Bcr-related; AC, anthracycline; Bcr, breakpoint cluster region; CAD, coronary artery disease; CMP, cardiomyopathy; CV, cardiovascular; HER2, human epidermal growth factor receptor 2; LVEF, left ventricular ejection fraction; PAD, peripheral artery disease; VEGF, vascular endothelial growth factor; VHD, valvular heart disease. Modified from Čelutkienė *et al.* (2020).

and maximum patient safety (Čelutkienė et al. 2020). Accordingly, three categories (low, medium, and high) of cardiotoxicity risk were identified on the basis of cardiovascular risk factors, pre-existing cardiovascular disease, and type and dose of cancer therapy (Table 8.2; Čelutkienė et al. 2020).

It is essential to evaluate cardiac function with echocardiography before starting potentially cardiotoxic therapy in every cancer patient as a baseline for monitoring and for risk stratification.

The authors then proposed a personalized echocardiographic surveillance according to cardiotoxicity risk stratification in patients under anthracyclines and/or HER2-targeted treatment, during and after chemotherapy (Tables 8.3 and 8.4).

Table 8.3 Echocardiographic surveillance during and after anthracycline chemotherapy.

Baseline cardiotoxicity risk	During chemotherapy	Following chemotherapy
Low	• • Baseline • Following cycle completing cumulative lifetime dose of 240 mg/m^2 doxorubicin or equivalent[a] • Every additional 100 mg/m^2 doxorubicin above 240 mg/m^2 or every two cycles	• 12 months after final cycle • 5 yearly review
Medium	• Baseline • Following 50% of planned total treatment or every two cycles (optional) • Following cycle completing cumulative lifetime cycle of 240 mg/m^2 doxorubicin or equivalent[a]	• 12 months after final cycle • 5 yearly review
High	• Baseline • Every two cycles • Consider after every cycle above 240 mg/m^2 doxorubicin or equivalent[b]	• 6 months after final cycle[c] • 12 months after final cycle • Annually for 2 or 3 years thereafter, and then in 3- to 5-year intervals for life

NB. All low and medium cardiovascular risk cancer patients who develop new cardiac symptoms or new left ventricular dysfunction during treatment are reclassified as high cardiovascular risk and if chemotherapy continues, they should follow the high-risk surveillance.

[a]240 mg/m^2 doxorubicin is equivalent to 360 mg/m^2 epirubicin, 320 mg/m^2 daunorubicin, and 50 mg/m^2 idarubicin.

[b]300 mg/m^2 doxorubicin is equivalent to 420 mg/m^2 epirubicin, 400 mg/m^2 daunorubicin, and 60 mg/m^2 idarubicin.

[c]Depending upon symptoms and evidence of new left ventricular dysfunction during treatment.

Modified from Čelutkienė *et al.* (2020).

In patients under vascular endothelial growth factor inhibitor (VEGFi) and Bcr-Abl tyrosine kinase inhibitor (TKi) treatment, the authors suggested echocardiography every 4 months during the first year of treatment with an additional early assessment 2–4 weeks after starting treatment in patients with high baseline CV risk.

In long-term treatment with VEGFi and second- and third-generation Bcr-Abl TKi at 6–12 months, echocardiography should be considered.

In patients who are candidates for dasatinib, pre-treatment echocardiography screening to assess for pre-existing pulmonary hypertension is recommended as well as maintaining a low threshold for repeating echocardiography if cardiac symptoms develop. The decision to stop the treatment if new pulmonary arterial hypertension is detected may require right heart catheterization.

Table 8.4 Echocardiographic surveillance during and after HER2-targeted therapies.

Baseline cardiotoxicity risk	During chemotherapy	Following chemotherapy
Early invasive HER2+ breast cancer with neoadjuvant or adjuvant trastuzumab[a]		
Low	• Baseline • Every four cycles	• Optional 6–12 months after final cycle
Medium	• Baseline • Every three cycles, then every four if stable at 4 months[c]	• 6 months after final cycle • Optional 12months after final cycle
High	• Baseline • Every two cycles, then every three if stable at 3 months[d]	• 3 and 12months after final cycle • Optional 6 months after final cycle
Metastatic HER2+ breast cancer or gastric cancer with long-term HER2-targeted therapies[b]		
Low	• Baseline • Every four cycles in year 1 and every six cycles in year 2, then six monthly	Not indicated unless symptomatic
Medium	• Baseline • Every three cycles, then if stable six monthly[c]	Not indicated unless symptomatic
High	• Baseline • Every two or three cycles for 3 months, every four cycles in year 1, then reduce frequency[d]	Not indicated unless symptomatic

[a]Neoadjuvant trastuzumab or trastuzumab and pertuzumab.
[b]Long-term trastuzumab, trastuzumab and pertuzumab, or trastuzumab emtansine.
[c]Choice of 2 or 3 depends upon variables including baseline left ventricular function, cardiovascular history, baseline troponin, and previous anthracycline chemotherapy. In patients starting with surveillance after the first two cycles, reducing to every three and then every four from 6–12 months (and thereafter in metastatic patients) if asymptomatic and left ventricular function stable is recommended.
[d]In high-risk patients, close surveillance every two cycles is recommended for the first four cycles and then reducing to every three cycles for the remainder of the first year of treatment.
Modified from Čelutkienė *et al.* (2020).

Given the high cardiovascular events rate, baseline echocardiography should be performed in patients with myeloma multiple scheduled to receive proteasome inhibitors (bortezomib, carfilzomib, and ixazomib) which also allows assessment for cardiac AL amyloidosis. Follow-up echocardiography is suggested in medium–high risk patients receiving carfilzomib, and in every case of cardiac symptoms and signs development.

CV toxicity associated with ICI (e.g., ipilimumab, nivolumab, pembrolizumab, atezolizumab, avelumab, and durvalumab), including myocarditis sometimes causing cardiogenic shock and non-inflammatory LV systolic dysfunction, was initially considered rare (<1%), but with expanding use, its incidence is increasing. A definite strategy of echocardiographic surveillance in patients under ICI has not been established; however, serial echocardiographic screening may be considered in patients at high risk (combination ICI, ICI in combination with a second oncology drug with known cardiotoxicity, and significant pre-existing cardiovascular disease). The echocardiographic findings may vary from a normal examination to reduced wall thickening, reduced GLS, regional and global wall motion abnormalities, and/or diastolic dysfunction. Particularly, a reduction in GLS is an early sign of ICI-induced myocarditis. The timing and duration of surveillance remains to be determined as severe myocarditis and pericarditis usually appear early (within the first four cycles), whereas non-inflammatory LV dysfunction emerges later.

8.6 Other imaging modalities

8.6.1 Cardiac Magnetic Resonance

The routine use of CMR in cardio-oncology for surveillance is not feasible due to the lack of widespread accessibility and relatively high cost. However, when available, it is a very useful tool to identify changes in ventricular volumes and EF, especially in patients with poor quality echocardiographic images (Pepe et al. 2016), if a discrepancy between measurements of LV function exists or if myocardial perfusion assessment for ischemia is simultaneously planned.

Indeed, CMR is the gold standard for LV and RV volumes and function assessment. The standard CMR approach for quantifying biventricular function parameters uses contiguous short-axis slices covering the entire ventricles acquired from a cine steady-state free precession (SSFP) sequence. Armstrong *et al.* (2012), in a cohort of long-term survivors of chemotherapy, showed that 3DE was superior to 2DE, but both 3DE and 2DE were suboptimal at identifying patients with LVEFs below a threshold value of 50% defined by CMR. Cardiotoxicity may also involve the RV, and CMR is the most accurate method for its morphological and functional assessment because the irregular shape of the RV cannot be derived by geometrical formulas and requires 3D acquisition.

Moreover, it has been recognized that prior anthracycline exposure is associated with future declines in LV mass (Neilan et al. 2012). Potential etiologies for this decline include myocellular injury or death with or without the occurrence of intracellular fibrosis or myocardial progenitor cell population depletion (Drafts et al. 2013). For this purpose, CMR is the best technique, allowing an accurate and reproducible LV mass quantification, superior to echocardiography.

Myocardial tissue characterization represents an additive and incremental value of CMR in detecting cardiotoxicity. Chemotherapeutic agents, in fact, can cause edema and hyperemia and even cellular necrosis and subsequent fibrosis. Two small studies have demonstrated the early presence of myocardial edema by CMR in patients treated with cardiotoxic chemotherapy (Oberholzer et al. 2004; Grover et al. 2013).

Regardless, no prognostic data are available regarding the detection of edema by CMR after chemotherapy.

Through late gadolinium enhancement (LGE), CMR identifies areas of necrosis/fibrosis. Some small studies failed to detect LGE in chemotherapy-treated patients (Drafts et al. 2013; Ylänen et al. 2013), while others did not find a relevant prognostic role of LGE in cancer patients (Fallah-Rad et al. 2008; Wadhwa et al. 2009). Thus, larger studies with longer follow-up are needed for an understanding of the prevalence and prognostic impact of LGE in monitoring CT. However, LGE is able to detect only macroscopic fibrosis, while chemotherapy-treated patients can develop mild diffuse myocardial fibrosis that can be detected by T1 mapping with the evaluation of the extracellular volume. Some small studies have reported encouraging results on the use of T1 mapping in chemotherapy-related cardiac damage. T1 mapping represents a very useful technique that is able to detect changes in the molecular features of the myocardium prior to the occurrence of functional alterations. In this context, CMR is particularly important for cancer patients receiving ICI with new cardiac symptoms, arrhythmias, or cardiac troponin elevation when ICI-mediated myocarditis is suspected. Moreover recent data suggest that novel CMR indices may be potentially the earliest markers of AC-induced damage: an intracellular water life time τic, related to the size of cardiomyocyte, and a prolongation of T2 relaxation time, correlated with intra-cardiomyocyte edema (Čelutkienė et al. 2020). Nevertheless, its role in the detection of the early cardiotoxicity should be further investigated to provide a prognostic role for these findings and to ascertain whether this technique can improve the management of patients.

The limited availability, high cost, and time-consumption compared to echocardiography do not allow widespread use of CMR in cardiotoxicity surveillance. Thus, the use of CMR should be encouraged in situations in which discontinuation of chemotherapeutic regimens secondary to CTRCD is being entertained or when, because of technical limitations or the quality of echocardiographic images, the estimation of the LVEF is thought to be unreliable. Finally, issues with claustrophobia and hazards associated with ferromagnetic devices in some patients with cancer (e.g., breast tissue expanders used for breast reconstruction after mastectomy) must be considered.

8.7 Cardiac Nuclear Imaging

Radionuclide angiography (MUGA), due to its accuracy and reproducibility in LVEF measurement, was applied in academic centers, community hospitals, and physicians' offices for monitoring anthracyclines-induced cardiotoxicity (Schwartz et al. 1987). However the radiation exposure limits its widespread use. MUGA also does not provide comprehensive information on right ventricle function, left and right atrial sizes, and the presence or absence of valvular or pericardial disease. Therefore, it is frequently used as a complementary or alternative technique to echocardiography and CMR, when not available.

Of note, the use of different imaging technique for cardiotoxicity surveillance in the same patient should be discouraged, due to differences in LV volumes and EF measurements among techniques. Therefore, the choice of a single tool for the serial monitoring of LVEF during chemotherapy is preferred.

In a retrospective study of Hodgkin's lymphoma patients receiving AC-containing chemotherapy, serial [18F] fluorodeoxyglucose (18F-FDG) positron emission tomography-CT scans showed an increase in cardiac FDG uptake, which was associated with a decline in LVEF (Sarocchi et al. 2018). (61) An increase in myocardial 18-fluorodeoxyglucose uptake is associated with LVEF decline in Hodgkin lymphoma patients treated with anthracycline. Increased myocardial glucose utilization has also been observed after trastuzumab and radiation therapy, probably linked to myocardial inflammation and cell damage. Given the common use of 18F-FDG PET to monitor cancer progression, this phenomenon of elevated 18F-FDG uptake might be exploited for cardiotoxicity surveillance. Cardiac FDG-PET can be used to assess for ICI-mediated myocarditis in cases where CMR is not available, contraindicated, or provides equivocal results (Bonaca et al. 2019). There are also indications for nuclear imaging studies where a specific tracer can evaluate for the presence of cardiac metastases, for example, radiolabeled octreotide for cardiac carcinoid metastases.

8.8 Multimodality Imaging in Screening and Follow-up in Radiotherapy

Although the magnitude of the risk of "radiation-induced" heart disease (RIHD) with modern radiotherapy techniques is not yet well defined, screening and follow-up examinations are warranted. The manifestations of RIHD may acutely develop but most often become clinically apparent several years after irradiation. RIHD includes a wide range of deleterious effects on the heart: pericarditis, CAD, valvular heart disease, rhythm abnormalities, and non-ischemic myocardial and conduction system damages.

A cardiovascular screening for risk factors and a careful clinical examination should be performed in all patients. Younger age, cardiovascular risk factors or pre-existing cardiovascular diseases, exposure to high doses of radiation (>30 Gy), concomitant chemotherapy, anterior or left chest irradiation location (Hodgkin's lymphoma > left-sided breast cancer > right-sided breast cancer), and the absence of shielding designate highest risk and such patients are likely to benefit most from screening (Lancilotti et al. 2013).

A baseline comprehensive echocardiographic evaluation is warranted in all patients before initiating the radiotherapy. To assess cardiac structural and functional changes after radiation exposure, available techniques such as echocardiography, CMR, cardiac computed tomography (CT), or SPECT should be used meaningfully within the appropriate clinical indication.

8.8.1 Pericardial disease

Echocardiography is the technique of choice in patients with suspected or confirmed pericardial disease. Although it is the first-line imaging modality in diagnosis and follow-up of effusive or constrictive pericarditis, CT and CMR are more sensitive techniques in the detection of specific anatomical abnormalities, such as pericardial thickening and calcification.

8.8.2 Myocardial dysfunction

As for chemotherapy-related cardiotoxicity monitoring, echocardiography is the first-line imaging modality to assess LV systolic and diastolic function.

CMR could help in case of suboptimal echocardiographic image quality or to provide additional structural information.

8.8.3 Valvular heart disease (see also above)

Radiotherapy can be responsible for mild left-sided valve regurgitation in the first 10 years post-radiation. Hemodynamically significant (≥ moderate valve disease) is more common >10 years following radiation. Current guidelines recommend echocardiographic surveillance of valvular heart disease (Plana et al. 2014; Lancilotti et al. 2013).

8.8.4 Coronary artery disease

The time interval for the development of significant CAD is ~5–10 years post-radiation (King et al. 1996; Hull et al. 2003).

Cancer survivors have a more rapid progression of pre-existing atherosclerosis, indicating a potential need for earlier and more aggressive approach in older patients with known coronary artery disease or risk factors. Conversely, in younger cancer survivors, a specific radiation-induced coronary disease, which is different from atherosclerosis, may develop following exposure to high radiation doses. Image-based stress testing, such as stress echocardiography, perfusion SPECT, and CMR, is indicated in irradiated patients who are symptomatic for angina or who develop new resting regional wall-motion abnormalities on a follow-up echocardiogram. CMR is able to directly image epicardial coronary artery stenosis, microvasculature on myocardial perfusion, ventricular function, and viability. With the advent of fast and reliable coronary artery imaging with cardiac CT, CMR is relegated to clinical assessment in younger patients for entities such as anomalous coronary vessels. Because of the high negative predictive value and the inability to assess the hemodynamic significance of detected obstructions, coronary CT angiography is mostly used to rule out the presence of CAD. Therefore, the role of surveillance CT to detect subclinical CAD has been proposed. As in the general population, in RT survivors, the accuracy of CT and calcium score in the diagnosis of significant CAD is high and demonstrates excellent negative predictive value. Moreover, recent data show that the inclusion of CT in the diagnostic workup of stable patients improves long-term prognosis by reducing the incidence of myocardial infarction (Newby et al. 2018).

Incidental coronary calcium in thoracic CT for staging and/or RT planning, subsequent follow-up CT and/or PET-CT scans should be reported and quantified. Coronary artery calcification obtained from non-gated chest CT scans correlates well with a 3-mm coronary calcium scan and is incrementally associated with worse CV outcomes in cancer patients implicating timely prescription of preventive therapies.

However, the timing of CT for surveillance in asymptomatic cancer survivors following high-dose radiation to the chest is unknown and requires further study.

During follow-up, a yearly history and physical examination with close attention to symptoms and signs of heart disease is essential. According to current recommendations, in patients who remain asymptomatic, screening echocardiography 10 years after treatment appears reasonable (Lancilotti et al. 2013). In cases where there are no pre-existing cardiac abnormalities, surveillance transthoracic echocardiogram should be then performed every 5 years.

In high-risk asymptomatic patients (patients who underwent anterior or left-side chest irradiation with ≥1 risk factors for RIHD), a screening echocardiography may be advocated after 5 years (Lancilotti et al. 2013). In these patients, the increased risk of coronary events 5–10 years after radiotherapy makes it reasonable to consider non-invasive stress imaging to screen for obstructive CAD. Repeated stress testing can be planned every 5 years if the first exam does not show inducible ischemia.

The additional role of CMR or cardiac CT depends on the initial echocardiographic results and the clinical indication as well as the local expertise and facilities. However, when the echocardiographic examination yields equivocal findings, these imaging modalities should be considered.

8.9 Conclusions and Future Directions

Cardiovascular imaging modalities play a key role in the developing field of cardio-oncology, providing highly sensitive methods for a timely diagnosis of cardiotoxicity.

Ideally each patient who must undergo chemotherapy and/or radiotherapy should have a baseline assessment through cardiac imaging. Echocardiography is the modality of choice and 3D echocardiography should be preferred when available for the calculation of volumes and EF. The first examination should be as complete as possible and must include the study of: systolic function, by LVEF and also GLS, diastolic function, heart valves, pericardium, and right chambers. Myocardial deformation imaging and 3D volumetric analysis seem to be optimal techniques to address temporal structural and functional changes during cancer therapy. If the quality of the echocardiogram is suboptimal, CMR is recommended.

Later checks ought to have to be guided by the results of the baseline examination, by the patient's cardiotoxicity risk and by cardiac biomarkers.

Moreover, patients' surveillance requires collaborative evaluation by the cardio-oncology team. Suggested detailed algorithms for anthracycline and HER2-targeted therapies aim to improve current clinical practice. Further studies are needed to establish if effective surveillance schemes may change the outcomes of oncology patients by improving their mortality and morbidity.

References

1. Agrawal V, Christenson ES, Showel MM. Tyrosine kinase inhibitor induced isolated pericardial effusion. Case Rep Oncol. 2015;8:88–93.
2. Armstrong GT, Plana JC, Zhang N, et al. Screening adult survivors of childhood cancer for cardiomyopathy: comparison of echocardiography and cardiac magnetic resonance imaging. J Clin Oncol. 2012;30:2876–84.
3. Badano LP, Aruta P, Nguyen K, et al. Principali applicazioni dell'ecocardiografia tridimensionale nell'attuale pratica clinica [Current clinical applications of three-dimensional echocardiography]. G Ital Cardiol (Rome). 2019;20:722–735.
4. Bonaca MP, Olenchock BA, Salem JE, et al. Myocarditis in the Setting of Cancer Therapeutics: Proposed Case Definitions for Emerging Clinical Syndromes in Cardio-Oncology. Circulation. 2019;140:80–91.
5. Çalık AN, Çeliker E, Velibey Y, et al. Initial dose effect of 5-fluorouracil: rapidly improving severe, acute toxic myopericarditis. Am J Emerg Med. 2012;30:257.e1–3.
6. Casey DJ, Kim AY, Olszewski AJ. Progressive pericardial effusion during chemotherapy for advanced Hodgkin lymphoma. Am J Hematol. 2012;87:521–4.
7. Čelutkienė J, Pudil R, López-Fernández T, et al. Role of cardiovascular imaging in cancer patients receiving cardiotoxic therapies: a position statement on behalf of the Heart Failure Association (HFA), the European Association of Cardiovascular Imaging (EACVI) and the Cardio-Oncology Council of the European Society of Cardiology (ESC). Eur J Heart Fail. 2020;22:1504–1524.
8. Charbonnel C, Convers-Domart R, Rigaudeau S, et al. Assessment of global longitudinal strain at low-dose anthracycline-based chemotherapy, for the prediction of subsequent cardiotoxicity. Eur Heart J Cardiovasc Imaging. 2017;18:392–401.
9. Cosyns B, Plein S, Nihoyanopoulos P, et al.; European Association of Cardiovascular Imaging (EACVI); European Society of Cardiology Working Group (ESC WG) on Myocardial and Pericardial diseases. European Association of Cardiovascular Imaging (EACVI) position

paper: Multimodality imaging in pericardial disease. Eur Heart J Cardiovasc Imaging. 2015;16:12–31.
10. Cosyns B, Plein S, Nihoyanopoulos P, et al.; European Association of Cardiovascular Imaging (EACVI); European Society of Cardiology Working Group (ESC WG) on Myocardial and Pericardial diseases. European Association of Cardiovascular Imaging (EACVI) position paper: Multimodality imaging in pericardial disease. Eur Heart J Cardiovasc Imaging. 2015;16:12–31.
11. Crawford MH. Chemotherapy-Induced Valvular Heart Disease. JACC Cardiovasc Imaging. 2016;9:240–2.
12. Di Lisi D, Bonura F, Macaione F, et al. Chemotherapy-induced cardiotoxicity: role of the conventional echocardiography and the tissue Doppler. Minerva Cardioangiol. 2011;59:301–8.
13. Dodos F, Halbsguth T, Erdmann E, et al. Usefulness of myocardial performance index and biochemical markers for early detection of anthracycline-induced cardiotoxicity in adults. Clin Res Cardiol. 2008;97:318–26.
14. Dogan SE, Mizrak D, Alkan A, et al Docetaxel-induced pericardial effusion. J Oncol Pharm Pract. 2017;23:389–391.
15. Drafts BC, Twomley KM, D'Agostino R Jr, et al. Low to moderate dose anthracycline-based chemotherapy is associated with early noninvasive imaging evidence of subclinical cardiovascular disease. JACC Cardiovasc Imaging. 2013;6:877–85.
16. Edoute Y, Haim N, Rinkevich D, et al. Cardiac valvular vegetations in cancer patients: a prospective echocardiographic study of 200 patients. Am J Med. 1997;102:252–8.
17. Eiken PW, Edwards WD, Tazelaar HD, et al. Surgical pathology of nonbacterial thrombotic endocarditis in 30 patients, 1985–2000. Mayo Clin Proc. 2001;76:1204–12.
18. Erven K, Florian A, Slagmolen P, et al. Subclinical cardiotoxicity detected by strain rate imaging up to 14 months after breast radiation therapy. Int J Radiat Oncol Biol Phys. 2013;85:1172–8.
19. Ewer MS, Lenihan DJ. Left ventricular ejection fraction and cardiotoxicity: is our ear really to the ground? J Clin Oncol. 2008;26:1201–3.
20. Fallah-Rad N, Lytwyn M, Fang T, et al. Delayed contrast enhancement cardiac magnetic resonance imaging in trastuzumab induced cardiomyopathy. J Cardiovasc Magn Reson. 2008;10:5.
21. Gaya AM, Ashford RF. Cardiac complications of radiation therapy. Clin Oncol (R Coll Radiol). 2005;17:153–9.

22. Grover S, Leong DP, Chakrabarty A, et al. Left and right ventricular effects of anthracycline and trastuzumab chemotherapy: a prospective study using novel cardiac imaging and biochemical markers. Int J Cardiol. 2013;168:5465–7.
23. Hare JL, Brown JK, Leano R, et al. Use of myocardial deformation imaging to detect preclinical myocardial dysfunction before conventional measures in patients undergoing breast cancer treatment with trastuzumab. Am Heart J. 2009;158:294–301.
24. Huang SY, Chang CS, Tang JL, et al. Acute and chronic arsenic poisoning associated with treatment of acute promyelocytic leukaemia. Br J Haematol. 1998;103:1092–5.
25. Hull MC, Morris CG, Pepine CJ, et al. Valvular dysfunction and carotid, subclavian, and coronary artery disease in survivors of hodgkin lymphoma treated with radiation therapy. JAMA. 2003;290:2831–7.
26. Jenkins C, Moir S, Chan J, et al. Left ventricular volume measurement with echocardiography: a comparison of left ventricular opacification, three-dimensional echocardiography, or both with magnetic resonance imaging. Eur Heart J. 2009;30:98–106.
27. Jurcut R, Wildiers H, Ganame J, et al. Strain rate imaging detects early cardiac effects of pegylated liposomal Doxorubicin as adjuvant therapy in elderly patients with breast cancer. J Am Soc Echocardiogr. 2008;21:1283–9.
28. Kang Y, Cheng L, Li L, et al. Early detection of anthracycline-induced cardiotoxicity using two-dimensional speckle tracking echocardiography. Cardiol J. 2013;20:592–9.
29. Katayama M, Imai Y, Hashimoto H, et al. Fulminant fatal cardiotoxicity following cyclophosphamide therapy. J Cardiol. 2009;54:330–4.
30. King V, Constine LS, Clark D, et al. Symptomatic coronary artery disease after mantle irradiation for Hodgkin's disease. Int J Radiat Oncol Biol Phys. 1996;36:881–9.
31. Klein AL, Abbara S, Agler DA, et al. American Society of Echocardiography clinical recommendations for multimodality cardiovascular imaging of patients with pericardial disease: endorsed by the Society for Cardiovascular Magnetic Resonance and Society of Cardiovascular Computed Tomography. J Am Soc Echocardiogr. 2013;26:965–1012.e15.
32. Kraigher-Krainer E, Shah AM, Gupta DK, et al.; PARAMOUNT Investigators. Impaired systolic function by strain imaging in heart failure with preserved ejection fraction. J Am Coll Cardiol. 2014;63:447–56. Erratum in: J Am Coll Cardiol. 2014;64:335.

33. Krupicka J, Marková J, Pohlreich D, et al.; German Hodgkin's Lymphoma Study Group. Echocardiographic evaluation of acute cardiotoxicity in the treatment of Hodgkin disease according to the German Hodgkin's Lymphoma Study Group. Leuk Lymphoma. 2002;43:2325–9.
34. Lancellotti P, Nkomo VT, Badano LP, et al.; European Society of Cardiology Working Groups on Nuclear Cardiology and Cardiac Computed Tomography and Cardiovascular Magnetic Resonance; American Society of Nuclear Cardiology, Society for Cardiovascular Magnetic Resonance, and Society of Cardiovascular Computed Tomography. Expert consensus for multi-modality imaging evaluation of cardiovascular complications of radiotherapy in adults: a report from the European Association of Cardiovascular Imaging and the American Society of Echocardiography. J Am Soc Echocardiogr. 2013;26:1013–32. Erratum in: J Am Soc Echocardiogr. 2013;26:1305.
35. Lang RM, Badano LP, Mor-Avi V, et al. Recommendations for cardiac chamber quantification by echocardiography in adults: an update from the American Society of Echocardiography and the European Association of Cardiovascular Imaging. J Am Soc Echocardiogr. 2015;28:1–39.e14.
36. Lund MB, Ihlen H, Voss BM, et al. Increased risk of heart valve regurgitation after mediastinal radiation for Hodgkin's disease: an echocardiographic study. Heart. 1996;75:591–5.
37. Montani D, Bergot E, Günther S, et al. Pulmonary arterial hypertension in patients treated by dasatinib. Circulation. 2012;125:2128–37.
38. Mor-Avi V, Lang RM, Badano LP, et al. Current and evolving echocardiographic techniques for the quantitative evaluation of cardiac mechanics: ASE/EAE consensus statement on methodology and indications endorsed by the Japanese Society of Echocardiography. J Am Soc Echocardiogr. 2011;24:277–313.
39. Muraru D, Badano LP, Piccoli G, et al. Validation of a novel automated border-detection algorithm for rapid and accurate quantitation of left ventricular volumes based on three-dimensional echocardiography. Eur J Echocardiogr. 2010;11:359–68.
40. Murbraech K, Wethal T, Smeland KB, et al. Valvular Dysfunction in Lymphoma Survivors Treated With Autologous Stem Cell Transplantation: A National Cross-Sectional Study. JACC Cardiovasc Imaging. 2016;9:230–9.
41. Narayan HK, Finkelman B, French B, et al. Detailed Echocardiographic Phenotyping in Breast Cancer Patients: Associations With Ejection Fraction Decline, Recovery, and Heart Failure Symptoms Over 3 Years of Follow-Up. Circulation. 2017;135:1397–412.

42. Negishi K, Negishi T, Hare JLet al. Independent and incremental value of deformation indices for prediction of trastuzumab-induced cardiotoxicity. J Am Soc Echocardiogr. 2013;26:493–8.
43. Negishi T, Thavendiranathan P, Negishi K, et al.; SUCCOUR investigators. Rationale and Design of the Strain Surveillance of Chemotherapy for Improving Cardiovascular Outcomes: The SUCCOUR Trial. JACC Cardiovasc Imaging. 2018;11:1098–1105.
44. Neilan TG, Coelho-Filho OR, Pena-Herrera D, et al. Left ventricular mass in patients with a cardiomyopathy after treatment with anthracyclines. Am J Cardiol. 2012;110:1679–86.
45. Oberholzer K, Kunz RP, Dittrich M, et al Anthrazyklin-induzierte Kardiotoxizität: MRT des Herzens bei Kindern und Jugendlichen mit malignen Erkrankungen [Anthracycline-induced cardiotoxicity: cardiac MRI after treatment for childhood cancer]. Rofo. 2004;176:1245–50.
46. Pepe A, Pizzino F, Gargiulo P, et al. Cardiovascular imaging in the diagnosis and monitoring of cardiotoxicity: cardiovascular magnetic resonance and nuclear cardiology. J Cardiovasc Med (Hagerstown). 2016;17S1:e45-e54.
47. Pizzino F, Vizzari G, Qamar R, et al. Multimodality Imaging in Cardiooncology. J Oncol. 2015;2015:263950.
48. Plana JC, Galderisi M, Barac A, et al. Expert consensus for multimodality imaging evaluation of adult patients during and after cancer therapy: a report from the American Society of Echocardiography and the European Association of Cardiovascular Imaging. Eur Heart J Cardiovasc Imaging. 2014;15:1063–93.
49. Poterucha JT, Kutty S, Lindquist RK, et al. Changes in left ventricular longitudinal strain with anthracycline chemotherapy in adolescents precede subsequent decreased left ventricular ejection fraction. J Am Soc Echocardiogr. 2012;25:733–40.
50. Santoro C, Esposito R, Lembo M, et al. Strain-oriented strategy for guiding cardioprotection initiation of breast cancer patients experiencing cardiac dysfunction. Eur Heart J Cardiovasc Imaging. 2019;20:1345–1352.
51. Sarocchi M, Bauckneht M, Arboscello E, et al. An increase in myocardial 18-fluorodeoxyglucose uptake is associated with left ventricular ejection fraction decline in Hodgkin lymphoma patients treated with anthracycline. J Transl Med. 2018;16:295.
52. Savoia F, Gaddoni G, Casadio C, et al. A case of aseptic pleuropericarditis in a patient with chronic plaque psoriasis under methotrexate therapy. Dermatol Online J. 2010;16:13.

53. Sawaya H, Sebag IA, Plana JC, et al. Assessment of echocardiography and biomarkers for the extended prediction of cardiotoxicity in patients treated with anthracyclines, taxanes, and trastuzumab. Circ Cardiovasc Imaging. 2012;5:596–603.
54. Sawaya H, Sebag IA, Plana JC, et al. Early detection and prediction of cardiotoxicity in chemotherapy-treated patients. Am J Cardiol. 2011;107:1375–80.
55. Schwartz RG, McKenzie WB, Alexander J, et al. Congestive heart failure and left ventricular dysfunction complicating doxorubicin therapy. Seven-year experience using serial radionuclide angiocardiography. Am J Med. 1987;82:1109–18.
56. SCOT-HEART Investigators, Newby DE, Adamson PD, Berry C, et al. Coronary CT Angiography and 5-Year Risk of Myocardial Infarction. N Engl J Med. 2018;379:924–933.
57. Stoodley PW, Richards DA, Hui R, et al. Two-dimensional myocardial strain imaging detects changes in left ventricular systolic function immediately after anthracycline chemotherapy. Eur J Echocardiogr. 2011;12:945–52.
58. Tassan-Mangina S, Codorean D, Metivier M, et al. Tissue Doppler imaging and conventional echocardiography after anthracycline treatment in adults: early and late alterations of left ventricular function during a prospective study. Eur J Echocardiogr. 2006;7:141–6.
59. Toro-Salazar OH, Ferranti J, Lorenzoni R, et al. Feasibility of Echocardiographic Techniques to Detect Subclinical Cancer Therapeutics-Related Cardiac Dysfunction among High-Dose Patients When Compared with Cardiac Magnetic Resonance Imaging. J Am Soc Echocardiogr.;29:119–31.
60. van Nimwegen FA, Schaapveld M, Janus CP, et al. Cardiovascular disease after Hodgkin lymphoma treatment: 40-year disease risk. JAMA Intern Med. 2015;175:1007–17.
61. Wadhwa D, Fallah-Rad N, Grenier D, et al. Trastuzumab mediated cardiotoxicity in the setting of adjuvant chemotherapy for breast cancer: a retrospective study. Breast Cancer Res Treat. 2009;117:357–64.
62. Ylänen K, Poutanen T, Savikurki-Heikkilä P, et al. Cardiac magnetic resonance imaging in the evaluation of the late effects of anthracyclines among long-term survivors of childhood cancer. J Am Coll Cardiol. 2013;61:1539–47.
63. Zamorano JL, Lancellotti P, Rodriguez Muñoz D, et al. 2016 ESC Position Paper on cancer treatments and cardiovascular toxicity developed under the auspices of the ESC Committee for Practice Guidelines: The Task

Force for cancer treatments and cardiovascular toxicity of the European Society of Cardiology (ESC). Eur Heart J. 2016;37:2768–801.

64. Zito C, Longobardo L, Cadeddu C, et al. Cardiovascular imaging in the diagnosis and monitoring of cardiotoxicity: role of echocardiography. J Cardiovasc Med (Hagerstown). 2016;17S1:e35-e44.
65. Zito C, Longobardo L, Citro R, et al. Ten Years of 2D Longitudinal Strain for Early Myocardial Dysfunction Detection: A Clinical Overview. Biomed Res Int. 2018;2018:8979407.
66. Zito C, Manganaro R, Cusmà Piccione M, Madonna R, Monte I, Novo G, Mercurio V, Longobardo L, Cadeddu Dessalvi C, Deidda M, Pagliaro P, Spallarossa P, Costantino R, Santarpia M, Altavilla G, Carerj S, Tocchetti CG. Anthracyclines and regional myocardial damage in breast cancer patients. A multicentre study from the Working Group on Drug Cardiotoxicity and Cardioprotection, Italian Society of Cardiology (SIC). Eur Heart J Cardiovasc Imaging. 2021 Mar 22;22(4):406–415.

9

Venous Thromboembolism in Cardio-Oncology

Ciro Santoro MD, PhD & Mario Enrico Canonico MD

Department of Advanced Biomedical Science, Federico II University Hospital, Naples, Italy

Correspondence to:
Dr. Ciro Santoro, MD, PhD
Dpt. Of Advanced Biomedical Science, Federico II University Hospital, Naples, Via Sergio Pansini 5, bld 1, 80131 Naples, Italy,
Tel: +39 081 7463663, Fax: +39 081 7464255 Email: ciro.santoro@unina.it

KEYWORDS: Anticoagulation Treatment; Prognosis; Risk Stratification; Thromboprophylaxis; Venous Thromboembolism.

9.1 Introduction

Acute venous thrombosis could unveil occult cancer being its first manifestation (Carrier et al. 2008). Approximately 20%–30% of all first venous thrombotic events are cancer related (White et al. 2005; Spencer et al. 2007). Furthermore, the presence of active malignancy is considered a potential factor for unfavorable evolution and proximal progression of distal deep vein thrombosis (DVT). Such risk orientates to a more aggressive anticoagulation management rather than a watchful ultrasound surveillance. In cancer patients, anticoagulation management not only should be aggressive but also last longer given the fact that the presence of cancer is considered a “non-transient” risk factor for venous thromboembolism (VTE), resulting in higher VTE recurrences risk (Ortel et al. 2020). On the other side, cancer may increase the bleeding risk as well, whereas hemorragic events account for higher mortality in oncologic population, occurring in about 10% of solid tumor and higher proportion in patients with hematologic malignancies

(Reeves and Key 2012). Consequently, the simultaneous presence of cancer and VTE put the clinician in a tough spot in terms of balancing the thrombotic and hemorragic risk.

9.2 Biological Mechanism of Cancer-Related Thrombosis

Malignancy interacts in an intricate way with hemostatic system enabling both thrombosis and hemorrhage. The main pathway involved in these reactions lies in the activation of tumor-associated clot promoting factors that finally increase the concentration of active thrombin and fibrin. This reaction induces mobilization of platelet, leukocyte, and endothelial cells which in return deploy surface adherence and procoagulant factors (Falaga et al. 2013). Biological mechanism beyond cancer-induced procoagulant status depends on different processes that induce a derangement of the hemostatic system. Cancer procoagulant (CP), highly expressed by proliferating blast cell and by some solid tumors, directly activates Factor X (Kowich et al. 1994; Molnar et al. 2009). Procoagulant tissue factor (TF), overexposed on tumor cell surface, could be vacuolized and released in tumor microparticles rich in TF, which disseminate into the body and causing both systemic and localized thrombotic events. Mice injected with microparticles enriched of TF showed a higher incidence of disseminated intravascular coagulation (DIC)-like syndrome, thus enforcing this pathophysiological hypothesis (Falanga et al. 2012). In addition, the production of pro-inflammatory cytokines, like TNF-α and IL-1β, and proangiogenic (mainly VEGF and FGF) by malignant cell further contribute to the procoagulant status by inducing cell-adhesion and vascular cells activation (Falanga et al. 2009) (Figure 9.1).

9.3 Epidemiology and Risk Stratification

The occurrence of cancer-related venous thrombosis is estimated to be between 20% and 30 % among all the venous thromboembolic events (Heit et al. 2002; White et al. 2005; Imberti et al. 2008; Braekkan et al. 2010; Gussoni et al. 2013). Recent or active malignant neoplasm increases the risk of venous thrombosis four- to seven-fold, compared to non-oncologic patients, according to the results of several international registries (Heit et al. 2000; Blom et al. 2005; Cronin-Fenton et al. 2010; Walker et al. 2013). In the Multiple Environmental and Genetic Assessment of risk factor for venous thrombosis (MEGA) study, more than 3000 patients experiencing DVT were included in the study. The occurrence of cancer in this population increased the risk of DVT by seven-fold (Blom et al. 2005).

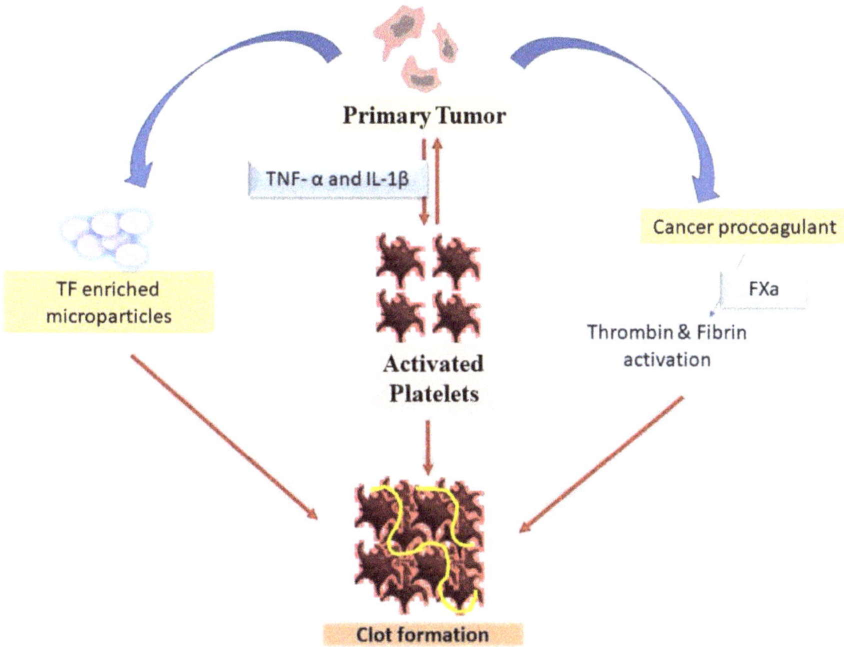

Figure 9.1 Synergic interaction between primary tumor and activated platelets to develop clot formation due to FX_a, thrombin, and TF release from cancer cells.

Furthermore, different cohort studies examined the absolute risk for DVT in cancer patients. The resulting cumulative incidence appears to have wide variation, due to several factors that might influence the observed event, such as time from cancer diagnosis, method of detection for DVT, and follow-up duration. When patients are observed from the time of cancer diagnosis, the cumulative risk for DVT may correspond to 1%–8% within 1–2 years (Sallah et al. 2002; Chew et al. 2006; Vormittag et al. 2009).

Over the last two decades, the incidence of venous thrombosis in patients with malignancies showed hyperbolic increment, with a 1.5% incidence in the late 1980s to 4.6% in the early 2000s (Khorana et al. 2007). This increment could be a consequence of different factors; first, the improvement in diagnostic testing and the spread acknowledges of solid link between cancer and venous thrombotic events. Second, the progressive longer survival of cancer population, due to the development of more effective multiple treatments that however held some degree of prothrombotic risk.

However, stratifying the risk of DVT in cancer patients according to their background risk may predict the incidence of DVT. In fact, patients with

Table 9.1 Risk factors for venous thrombosis.

Cancer-related	Patients-related
Type of cancer	Older age
Stage of cancer	Prolonged immobility
Time since cancer diagnosis	Prior history of venous thrombosis
Treatment	Black ethnicity
	Prothrombotic mutation
	Comorbidities (≥ 3):
	• Venous thrombosis
	• Pulmonary disease
	• Renal disease
	• Infection
	• Anemia

high grade or metastatic disease or the presence of prothrombotic anticancer treatment have a two-fold risk increment to develop DVT compared to cancer patients at average risk (Horsted et al. 2012).

The risk for developing venous thrombotic events in oncologic setting could depend on several cancer-related and patients-related factors (Table 9.1).

Type of malignancy seems to influence the risk of venous thrombosis to the point where high-risk cancer (pancreas, brain, lung, ovarian, lymphoma, myeloma, kidney, stomach, and bone cancer) and low-risk cancer (breast, prostate, malignant melanoma, and testicular) can be identified (Cronin-Fenton et al. 2010). Cancer features of biological aggressiveness, such as early metastatic diffusion and consequential dismal prognosis, correlated with the incidence of venous thrombosis (Wun and White 2009), as evidenced by Tim *et al.* who showed a positive association between incidence rates of venous thrombotic events plotted against 1-year relative mortality for different types of cancer (Timp et al. 2013). In a Danish cohort study, which included cancer patients at first diagnosis, 57,591 oncologic patients were enrolled and, among them, 1023 experienced VTE. In a subanalysis, the relative risk of thrombotic events progressively increased along with the tumor stage, with an adjusted relative risk (aRR) of 2.9 for stages I and II and an aRR of 7.5 and 17.1 for stages III and IV, respectively (Cronin-Fenton et al. 2010). Similarly, tumor grading appears also to play a role in the association of cancer and VTE as shown by the Vienna Cancer and Thrombosis Study, where 747 patients with solid tumors were investigated, looking for VTE occurrence. In this study, the presence of high-grade tumor (G3–G4) exposed the patients to a higher risk of VTE compared to those with low grade (G1–G2) (hazard ratio, 2.0; 95% CI, 1.1–3.5; $p = 0.015$).

Another debated risk factor for VTE in cancer is the higher incidence of thrombotic events in the first months after cancer diagnosis (Ahlbrecht et al.

2012). The determinants behind this phenomenon can be detected by observing the clinical evolution of the disease. In the early phases after cancer diagnosis, patients start cancer treatment, whose protocol frequently includes loading dosage that may determine prothrombotic mechanism. On the other hand, in the late phase, anticancer treatment may induce either cancer regression, when efficacious, or death, in the presence of unfortunate prognosis. In both of these cases, the incidence of DVT observed will inexorably fall. As mentioned above, treatment may paradoxically increase the occurrence of DVT, as is the case of hormonal therapy and its boosting effect on venous thrombotic risk, especially in breast cancer (Clahsen et al. 1994). In gastrointestinal cancer, the use of cisplatin containing regimen either alone or in combination with epirubicin revealed higher cumulative incidence of VTE (Starling et al. 2009). Recent chemotherapy protocol with monoclonal antibody (i.e., bevacizumab) (Nalluri et al. 2008) and immunomodulatory drugs (i.e., lenalidomide) (Hirsh 2007) are also factors that might increase the risk of venous thrombosis.

Supportive therapy for prophylaxis or treatment of anemia in cancer patients frequently associated with concurrent antineoplastic therapy regimen such as erythropoiesis-stimulating agents and red blood cell transfusion increase the risk of thrombo-embolic events as demonstrated in a meta-analysis that reviews 57 trials on cancer patients (Bohlius et al. 2006).

In the oncologic setting, besides cancer-related risk factor, patients-related risk factors too could play a role in increasing the risk for venous thrombotic events. The main patients-related risk factors in this setting are age, history of previous venous thrombosis, and comorbidities. Among the latter, arterial thromboembolism, pulmonary and renal disease, infection, and anemia showed to elevate the risk of thrombosis (Horsted et al. 2012). In patients with colorectal cancer, the concomitant presence of more than three chronic comorbidities condition increases the risk of venous thrombosis of two-fold within 1 year after the diagnosis (Alcalav et al. 2006). Other common risk factors for venous thrombosis in non-cancer population that may recur in this context, such as prolonged immobility after a surgical treatment, placement of central venous catheters, or prothrombotic mutations, should also be a concern as risk factors for thrombosis in cancer patients (Blom et al. 2005; Dentali et al. 2008).

9.4 Clinical Presentation

Cancer-related thrombotic events usually interest the venous side and manifest clinically in a broad spectrum of condition, also addressed as Trousseau's syndrome, named after the French physician who first described the association between thrombosis and cancer. This syndrome includes venous and arterial thrombosis, non-bacterial thrombotic endocarditis (NBTE), thrombotic microangiopathy (TMA), and veno-occlusive disease (VOD).

Venous thrombosis usually interests lower limbs and their clinical presentation in cancer patients may not differ from those patients without cancer (Figure 9.2).

Beyond that, cancer related DVT may manifest in several sites (ileocaval, portal or extrahepatic, mesenteric, upper limb veins, migratory superficial thrombophlebitis, etc.), whose singularity should call attention on possible unknown malignancies (Blom et al. 2006). Mieloprolipherative neoplasm (MPN), for instance, may first manifest through cerebral and splanchnic veins thrombosis (i.e., Budd–Chiari syndrome and portal vein thrombosis) (Reikvam and Tiu 2012). Furthermore, in hematologic malignancies, disarrangement of the microcirculatory system generates a broad spectrum of manifestations such as erythromelalgia, cerebrovascular disorder, and non-bacterial thrombotic endocarditis (NBTE). NBTE, frequently associated with MPN, can also be observed in solid tumors, occurring in up to 1.3% of patients dying of cancer. It embodies a cardiac manifestation of hemostatic

A

B

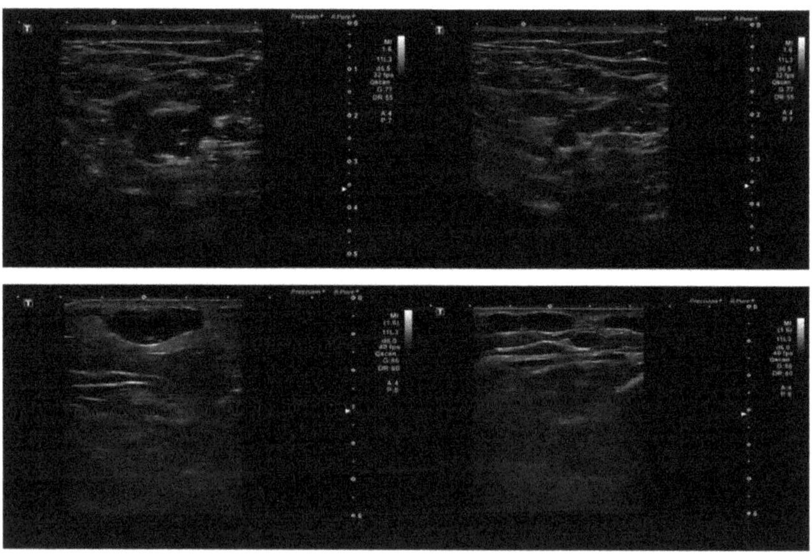

Figure 9.2 (**A**) Not-occlusive deep venous thrombosis of the right common femoral vein with no compression ultrasound (left panel) and with compression ultrasound (right panel, CUS positive, only partial compressibility). (**B**) On the left, superficial venous thrombosis of the right greater saphenous vein (CUS positive). On the right, superficial venous thrombosis of collaterals of the right greater saphenous vein.

derangement, as a result of platelets and fibrin aggregation on cardiac valves (Sanon et al. 2011), which could be responsible for arterial thromboembolic events because of vegetation shed displaced peripherally and causing acute vascular occlusions and ischemia.

Additionally, the presence of central catheter placement in the upper limb, used for therapeutic purpose, might partially explain the higher frequencies of upper limb DVT in cancer.

Data about arterial thromboembolic events are also present, despite limited, mainly affecting upper and lower extremities and cerebral vasculature (Arboiix 2000). The incidence of these events is estimated to be around 2%–5%, but it might vary depending on the type of tumor and chemotherapeutic protocol utilized.

Catastrophic systemic activation of the coagulation cascade may lead to disseminated intravascular coagulation (DIC) that will eventually lead to multi-organ failure and increased risk of major and fatal bleeding due to consumption of clotting factors and platelets. DIC is frequently associated with hematologic malignancies such as acute leukaemias (Sanz et al. 2009).

9.5 Recurrent Venous Thrombosis and Treatment in Cancer Population

Cancer patients held an elevated risk of recurrent venous thrombosis, with a two- to three-fold increase risk of recurrence compared with non-cancer patients (Prandoni et al. 1996, 2002; Trujillo-Santos et al. 2008). Of note, Prandoni *et al.* observed a higher incidence of recurrent venous thromboembolism despite the anticoagulation therapy in cancer patients compared to non-cancer population, along with higher risk of major bleeding. Recurrence and bleeding occurred predominantly during the initial management and primary anticoagulation treatment, being related with the extent and severity of cancer rather than anticoagulation treatment intensities outside the therapeutic range (Prandoni et al. 1996). Prediction scores like the Ottawa prognostic score have been developed to assess the risk of recurrence of venous thrombosis in cancer population. Based on four predictors (gender, primary tumor site, stage, and number of prior thrombotic events), the Ottawa score differentiates patient at low risk of recurrence (4%) from those at increased risk (16%) (Louzada et al. 2012).

The selection of patients who have a balanced risk-benefit profile for initiation of anticoagulation is complex, given individual patient goals and preferences, changing prognosis of specific cancers, common comorbidities, potential drug–drug interactions, underweight states, and competing risks of morbidity and mortality (Mosarla et al. 2019). It stands to reason that the occurrence of cancer affects heavily all the three phases of the anticoagulation management of DVT; initial management (consisting in the first 5–21 days), primary treatment (the 3–6 months that follow the initial management), and secondary prevention (whose duration depends on VTE recurrence) as classified by the recent Guidelines for management of venous thromboembolism, (Spencer et al. 2007).

In the pre-DOAC era, clinical trials in this population compared LMWH monotherapy with LMWH-bridged warfarin. International Society on Thrombosis and Haemostasis (ISTH) and National Comprehensive Cancer Network (NCCN) 2018 guidelines recommend LMWH as standard of care in treating cancer-associated VTE with fondaparinux and unfractionated heparin as alternative treatment options, compared to oral anticoagulant treatment with warfarin (Lee et al. 2003, 2013; Streiff et al. 2018). Furthermore, parental delivery has the advantage of avoiding gastrointestinal (GI) absorption that might be altered by emetogenic cancer treatment and or by surgical treatment.

2019 ASCO Clinical Practice Guideline Update (Key et al. 2018) underscored the importance of risk stratification for VTE risk and of effective treatment to reduce the risk of VTE recurrence and mortality. Of note, direct anticoagulants (DOACs) have been added as an option for both prophylaxis and treatment, based on the results of recent trials that compared LMWH with DOACs in cancer related VTE (Raskob et al. 2018; Young et al. 2018; Agnelli et al. 2020; McBane et al. 2020).

In both Hokusai VTE Cancer study and in Selected Cancer Patients at Risk of Recurrence of Venous Thromboembolism (SELECT-D) pilot study, the efficacy of edoxaban and rivaroxaban, respectively, appeared to be more effective in reducing VTE recurrence, at the expense of higher incidence of bleeding events when compared to LWMH. Those events seemed to be particularly higher in those cancer patients with GI malignancies. The efficacy of apixaban in reducing the risk of DVT recurrence was shown in the Caravaggio trial, in which 1155 cancer patients were randomized to receive monotherapy with either apixaban or daltaparin for 6 months. Apixaban showed to be non-inferior for the treatment of cancer associated venous thromboembolism in the absence of increased risk for major bleeding (Agnelli et al. 2020).

Moreover, the possibility of an "oral only" loading dosage during the initial management (rivaroxaban 15 mg BID for 21 days or apixaban 10 mg BID for 7 days) results in a facilitate management of anticoagulation therapy, avoiding the parental lead in and bridging with heparin.

Anticoagulant treatment DOACs provide a reasonable alternative to LMWH in the treatment of VTE in cancer patients; however, particular care should be spent to monitor bleeding events, especially gastrointestinal or urinary (Wojtukiewicz et al. 2020).

9.6 Prognosis

The concomitant presence of cancer and thromboembolic event reduces the survival rate of five-fold when compared to thrombotic event alone at 1 year (Heit et al. 2002). On the other hand, the survival rate of patients with cancer and venous thrombosis is more than halved at 1 year if compared to cancer population without venous thrombosis, after matching for age, gender, type of cancer, and year of diagnosis (Sørensen et al. 2000). Indeed, venous thrombosis results to be a significant prognosticator at 1 year for all cancer types, in a large cohort study of more than 230 thousand cancer patients, with and without thrombotic events (Chew et al. 2006). Even though the specific weight of thrombotic events alone on the mortality rates in cancer population is difficult to be estimated, thrombotic event gained the second place as cause of death in patients with cancer, after cancer progression (Khorana et al. 2007).

9.7 Thromboprophylaxis

Thromboprophylaxis for primary prevention of VTE has been an important recent clinical and research issue, and its benefit in cancer patients is largely debated. Despite a clear reduction in terms of incidence of venous thrombosis, the increment of major bleeding is a relevant deterrent for the generalization of this kind of management (Ay et al. 2009; Di Nisio et al. 2012). Thus, a possible approach could be to stratify the risk of thrombotic events, to frame those at increased risk of thrombotic events, who could benefit from a prophylactic anticoagulation. Different risk factors and bio-humoral markers have been proposed to increase prediction accuracy for thrombotic events (Iversen et al. 2002; Khorana et al. 2005, 2008; Kröger et al. 2006; Ay et al. 2009; Zwicker et al. 2009; Simanek et al. 2010) and specific efforts have been made to build a comprehensive prediction models to guide decision on thromboprophylaxis. Khorana *et al.* developed a risk model based on five variables (Table 9.2) assessed from a cohort study of 2701 cancer patients who had to start chemotherapy. Patients displaying a risk score higher than 3 (i.e., high risk group) before starting the treatment showed to have a 6.7% to 7.1% rates of developing venous thrombosis (Khorana et al. 2005).

Table 9.2 Predictive model for chemotherapy-associated venous thrombosis.

Patient characteristic	Risk score
Site of cancer	
Very high risk (stomach, pancreas, etc.)	2
High risk (lung, lymphoma, gynaecologic, bladder, testicular, etc.)	1
Prechemotherapy platelet count ≥ 350 × 100/L	1
Prechemotherapy hemoglobin level < 100 g/L or use of red cell growth factors	1
Prechemotherapy leukocyte count > 11 × 100/L	1
Body mass index ≥ 35 kg/m^2	1
Rates of venous thrombosis	**Incidence**
Low risk (score = 0)	0.8%–0.3%
Intermediate risk (score 1–2)	1.8%–2%
High risk (score ≥3)	6.7%–7.15%

(Khorana et al. 2005).

Furthermore, improvement in prediction of thrombotic events has been designed adding bio-humoral markers, (i.e., P-selectin >53.1 ng/mL and D-dimer levels >1.44 mg/mL) to the Khorana's risk model. This expanded model showed a sensitivity of 96% when the patients receive the point score of 1, thus reasonably excluding thromboprophylaxis, and a specificity of 98% for those at higher cutoff point (score ≥5), who may benefit from thromboprophylaxis (Ay et al. 2010). Limitation for this expanded score consists in scarce availability of those bio-humoral marker in daily routine.

9.8 Conclusions

Cancer-related venous thrombotic events are a not infrequent manifestation during the history of disease. The right choice between anticoagulation strategy, thrombo-hemorrhagic risk management, and patients' comorbidities represents a challenge for physicians. An accurate risk stratification to select patients at higher risk of thrombotic events, who would benefit from thromboprophylaxis, should be encouraged. Early identification and treatment of this complication is particularly relevant in the onco-hematologic setting, given the substantial impact of venous thrombotic events on morbidity and mortality. Despite increased risk of major bleeding, recently, DOACs provide an attractive alternative to LMWH in the treatment of VTE in cancer patients especially in those without drug interactions, impaired renal function, and low or high body mass index. The large amount of multiple connections between thrombotic pathway and cancer growth needs to be further elucidated in order to provide accurate prognostic score and targeted therapy.

References

1. Agnelli G, Becattini C, Meyer G, et al.; Caravaggio Investigators. Apixaban for the Treatment of Venous Thromboembolism Associated with Cancer. N Engl J Med. 2020;382:1599–1607.
2. Ahlbrecht J, Dickmann B, Ay C, et al. Tumor grade is associated with venous thromboembolism in patients with cancer: results from the Vienna Cancer and Thrombosis Study. J Clin Oncol. 2012;30:3870–3875.
3. Alcalay A, Wun T, Khatri V, et al. Venous thromboembolism in patients with colorectal cancer: incidence and effect on survival. J Clin Oncol. 2006;24:1112–1118.
4. Arboix A. [Cerebrovascular disease in the cancer patient]. Rev Neurol 2000;31:1250–1252
5. Ay C, Dunkler D, Marosi C, et al. Prediction of venous thromboembolism in cancer patients. Blood. 2010;116:5377–82.
6. Ay C, Simanek R, Vormittag R, et al. High plasma levels of soluble P-selectin are predictive of venous thromboembolism in cancer patients: results from the Vienna Cancer and Thrombosis Study (CATS). Blood. 2008;112:2703–2708.
7. Ay C, Vormittag R, Dunkler D, et al. D-dimer and prothrombin fragment 1 + 2 predict venous thromboembolism in patients with cancer: results from the Vienna Cancer and Thrombosis Study. J Clin Oncol. 2009;27:4124–4129.
8. Blom JW, Doggen CJ, Osanto S, et al. Malignancies, prothrombotic mutations, and the risk of venous thrombosis. JAMA. 2005;293: 715–722.
9. Blom JW, Vanderschoot JP, Oostindi¨er MJ, et al. Incidence of venous thrombosis in a large cohort of 66,329 cancer patients: results of a record linkage study. J Thromb Haemost. 2006;4:529–535.
10. Bohlius J, Wilson J, Seidenfeld J, et al. Recombinant human erythropoietins and cancer patients: updated meta-analysis of 57 studies including 9353 patients. J Natl Cancer Inst. 2006;98:708–714.
11. Braekkan SK, Borch KH, Mathiesen EB, et al. Body height and risk of venous thromboembolism: The Tromsø Study. Am J Epidemiol. 2010;171:1109–1115.
12. Carrier M, Le Gal G, Wells PS, et al. Systematic review: the Trousseau syndrome revisited: should we screen extensively for cancer in patients with venous thromboembolism? Ann Intern Med. 2008;149: 323–333.
13. Chew HK, Wun T, Harvey D, et al. Incidence of venous thromboembolism and its effect on survival among patients with common cancers. Arch Intern Med. 2006;166:458–464.

14. Clahsen PC, van de Velde CJ, Julien JP, et al. Thromboembolic complications after perioperative chemotherapy in women with early breast cancer: a European Organization for Research and Treatment of Cancer Breast Cancer Cooperative Group study. J Clin Oncol. 1994;12:1266–1271.
15. Cronin-Fenton DP, Søndergaard F, Pedersen LA, et al. Hospitalisation for venous thromboembolism in cancer patients and the general population: a population-based cohort study in Denmark, 1997–2006. Br J Cancer. 2010;103:947–953.
16. Dentali F, Gianni M, Agnelli G, et al. Association between inherited thrombophilic abnormalities and central venous catheter thrombosis in patients with cancer: a metaanalysis. J Thromb Haemost. 2008;6: 70–75.
17. Di Nisio M, Porreca E, Ferrante N, et al. Primary prophylaxis for venous thromboembolism in ambulatory cancer patients receiving chemotherapy. Cochrane Database Syst Rev. 2012:CD008500.
18. Falanga A, Marchetti M, Vignoli A. Coagulation and cancer: biological and clinical aspects. J Thromb Haemost. 2013;11:223–233.
19. Falanga A, Panova-NoevaM, Russo L. Procoagulantmechanisms in tumour cells. Best Pract Res Clin Haematol 2009;22:49–60.
20. Falanga A, Tartari CJ, Marchetti M. Microparticles in tumor progression. Thromb Res 2012;129(S1): S132–136.
21. Gussoni G, Frasson S, La Regina M, et al.; RIETE Investigators. Three-month mortality rate and clinical predictors in patients with venous thromboembolism and cancer. Findings from the RIETE registry. Thromb Res. 2013;131:24–30.
22. Heit JA, O'Fallon WM, Petterson TM, et al. Relative impact of risk factors for deep vein thrombosis and pulmonary embolism: a population-based study. Arch Intern Med. 2002;162:1245–1248.
23. Heit JA, Silverstein MD, Mohr DN, et al. Risk factors for deep vein thrombosis and pulmonary embolism: a population-based case-control study. Arch Intern Med. 2000;160:809–815.
24. Hirsh J. Risk of thrombosis with lenalidomide and its prevention with aspirin. Chest. 2007;131:275–277.
25. Horsted F, West J, Grainge MJ. Risk of venous thromboembolism in patients with cancer: a systematic review and meta-analysis. PLoS Med. 2012;9:e1001275.
26. Imberti D, Agnelli G, Ageno W, et al.; MASTER Investigators. Clinical characteristics and management of cancer-associated acute venous thromboembolism: findings from the MASTER Registry. Haematologica. 2008;93:273–278.

27. Iversen LH, Thorlacius-Ussing O. Relationship of coagulation test abnormalities to tumour burden and postoperative DVT in resected colorectal cancer. Thromb Haemost. 2002;87:402–408.
28. Key NS, Khorana AA, Kuderer NM, et al. Venous Thromboembolism Prophylaxis and Treatment in Patients With Cancer: ASCO Clinical Practice Guideline Update. J Clin Oncol. 2020;38:496–520.
29. Khorana AA, Francis CW, Culakova E, et al. Frequency, risk factors, and trends for venous thromboembolism among hospitalized cancer patients. Cancer. 2007;110:2339–2346.
30. Khorana AA, Francis CW, Culakova E, et al. Risk factors for chemotherapy-associated venous thromboembolism in a prospective observational study. Cancer. 2005;104: 2822–2829.
31. Khorana AA, Francis CW, Culakova E, et al. Thromboembolism is a leading cause of death in cancer patients receiving outpatient chemotherapy. J Thromb Haemost. 2007;5:632–634.
32. Khorana AA, Kuderer NM, Culakova E, et al. Development and validation of a predictive model for chemotherapy-associated thrombosis. Blood. 2008;111:4902–4907.
33. Kozwich DL, Kramer LC, Mielicki WP, et al. Application of cancer procoagulant as an early detection tumor marker. Cancer 1994;74:1367–1376.
34. Kröger K, Weiland D, Ose C, et al. Risk factors for venous thromboembolic events in cancer patients. Ann Oncol. 2006;17:297–303.
35. Lee AY, Bauersachs R, Janas MS, et al.; CATCH Investigators. CATCH: a randomised clinical trial comparing long-term tinzaparin versus warfarin for treatment of acute venous thromboembolism in cancer patients. BMC Cancer. 2013;13:284.
36. Lee AY, Levine MN, Baker RI, et al.; Randomized Comparison of Low-Molecular-Weight Heparin versus Oral Anticoagulant Therapy for the Prevention of Recurrent Venous Thromboembolism in Patients with Cancer (CLOT) Investigators. Low-molecular-weight heparin versus a coumarin for the prevention of recurrent venous thromboembolism in patients with cancer. N Engl J Med. 2003;349:146–153.
37. Louzada ML, Carrier M, Lazo-Langner A, et al. Development of a clinical prediction rule for risk stratification of recurrent venous thromboembolism in patients with cancer-associated venous thromboembolism. Circulation. 2012;126:448–454.
38. McBane RD 2nd, Wysokinski WE, Le-Rademacher JG, et al. Apixaban and dalteparin in active malignancy-associated venous thromboembolism: The ADAM VTE trial. J Thromb Haemost. 2020;18:411–421.

39. Molnar S, Guglielmone H, Lavarda M, et al. Procoagulant factors in patients with cancer. Hematology 2007;12:555–559.
40. Mosarla RC, Vaduganathan M, Qamar A, et al. Anticoagulation Strategies in Patients With Cancer: JACC Review Topic of the Week. J Am Coll Cardiol. 2019;73:1336–1349.
41. Nalluri SR, Chu D, Keresztes R, et al. Risk of venous thromboembolism with the angiogenesis inhibitor bevacizumab in cancer patients: a meta-analysis. JAMA. 2008;300: 2277–2285.
42. Ortel TL, Neumann I, Ageno W, et al. American Society of Hematology 2020 guidelines for management of venous thromboembolism: treatment of deep vein thrombosis and pulmonary embolism. Blood Adv. 2020;4:4693–4738.
43. Prandoni P, Lensing AW, Cogo A, et al. The long-term clinical course of acute deep venous thrombosis. Ann Intern Med. 1996;125:1–7.
44. Prandoni P, Lensing AW, Piccioli A, et al. Recurrent venous thromboembolism and bleeding complications during anticoagulant treatment in patients with cancer and venous thrombosis. Blood. 2002;100:3484–3488.
45. Raskob GE, van Es N, Verhamme P, et al.; Hokusai VTE Cancer Investigators. Edoxaban for the Treatment of Cancer-Associated Venous Thromboembolism. N Engl J Med. 2018;378:615–624.
46. Reeves BN, Key NS. Acquired hemophilia in malignancy. Thromb Res 2012; 129(S1): S66–68.
47. Reikvam H, Tiu RV. Venous thromboembolism in patients with essential thrombocythemia and polycythemia vera. Leukemia 2012;26:563–571
48. Sallah S, Wan JY, Nguyen NP. Venous thrombosis in patients with solid tumors: determination of frequency and characteristics. Thromb Haemost. 2002;87:575–579.
49. Sanon S, Lenihan DJ, Mouhayar E. Peripheral arterial ischemic events in cancer patients. Vasc Med 2011;16:119–30
50. Sanz MA, Grimwade D, Tallman MS, et al. Management of acute promyelocytic leukemia: recommendations from an expert panel on behalf of the European Leukemia Net. Blood 2009;113:1875–1891.
51. Simanek R, Vormittag R, Ay C, et al. High platelet count associated with venous thromboembolism in cancer patients: results from the Vienna Cancer and Thrombosis Study (CATS). J Thromb Haemost. 2010;8:114–20.
52. Sørensen HT, Mellemkjaer L, Olsen JH, et al. Prognosis of cancers associated with venous thromboembolism. N Engl J Med. 2000;343: 1846–1850.

53. Spencer FA, Lessard D, Emery C, et al. Venous thromboembolism in the outpatient setting. Arch Intern Med. 2007;167: 1471–1475.
54. Starling N, Rao S, Cunningham D, et al. Thromboembolism in patients with advanced gastroesophageal cancer treated with anthracycline, platinum, and fluoropyrimidine combination chemotherapy: a report from the UK National Cancer Research Institute Upper Gastrointestinal Clinical Studies Group. J Clin Oncol. 2009;27:3786–3793.
55. Streiff MB, Holmstrom B, Angelini D, et al. NCCN Guidelines Insights: Cancer-Associated Venous Thromboembolic Disease, Version 2.2018. J Natl Compr Canc Netw. 2018;16:1289–1303.
56. Timp JF, Braekkan SK, Versteeg HH, et al. Epidemiology of cancer-associated venous thrombosis. Blood. 2013;122:1712–1723
57. Trujillo-Santos J, Nieto JA, Tiberio G, et al.; RIETE Registry. Predicting recurrences or major bleeding in cancer patients with venous thromboembolism. Findings from the RIETE Registry. Thromb Haemost. 2008;100:435–439.
58. Vormittag R, Simanek R, Ay C, et al. High factor VIII levels independently predict venous thromboembolism in cancer patients: the cancer and thrombosis study. Arterioscler Thromb Vasc Biol. 2009;29:2176–2181.
59. Walker AJ, Card TR, West J, et al. Incidence of venous thromboembolism in patients with cancer - a cohort study using linked United Kingdom databases. Eur J Cancer. 2013;49:1404–1413.
60. White RH, Zhou H, Murin S, et al. Effect of ethnicity and gender on the incidence of venous thromboembolism in a diverse population in California in 1996. Thromb Haemost. 2005;93:298–305.
61. White RH, Zhou H, Murin S, et al. Effect of ethnicity and gender on the incidence of venous thromboembolism in a diverse population in California in 1996. Thromb Haemost. 2005;93:298–305.
62. Wojtukiewicz MZ, Skalij P, Tokajuk P, et al. Direct Oral Anticoagulants in Cancer Patients. Time for a Change in Paradigm. Cancers (Basel). 2020;12:1144.
63. Wun T, White RH. Epidemiology of cancer-related venous thromboembolism. Best Pract Res Clin Haematol. 2009;22:9–23.
64. Young AM, Marshall A, Thirlwall J, et al. Comparison of an Oral Factor Xa Inhibitor With Low Molecular Weight Heparin in Patients With Cancer With Venous Thromboembolism: Results of a Randomized Trial (SELECT-D). J Clin Oncol. 2018;36:2017–2023.
65. Zwicker JI, Liebman HA, Neuberg D, et al. Tumor-derived tissue factor-bearing microparticles are associated with venous thromboembolic events in malignancy. Clin Cancer Res. 2009;15:6830–6840.

Conclusions and Remarks

Valentina Mercurio, MD, PhD, FISC[1]; Pasquale Pagliaro, MD, PhD[2,3]; Claudia Penna, BSc, PhD[2,3]; Carlo G Tocchetti, MD, PhD, FHFA, FISC[1,4,5,6]

[1]Department of Translational Medical Sciences, Federico II University, Naples, Italy
[2]Departiment of Clinical and Biological Sciences, University of Turin, Torino, Italy
[3]Istituto Nazionale per le Ricerche Cardiovascolari, Bologna, Italy
[4]Center for Basic and Clinical Immunology Research (CISI), Federico II University, Naples, Italy
[5]Interdepartmental Center of Clinical and Translational Research (CIRCET), Federico II University, Naples, Italy
[6]Interdepartmental Hypertension Research Center (CIRIAPA), Federico II University, Naples, Italy

Cardio-oncology is an innovative and multidisciplinary field of contemporary medicine. Along with the increasing number of cancer survivors, cardiovascular complications due to cancer therapies are also growing. The data reported in the Pan-European CARDIOTOX-2020 registry underline the primary importance of this fascinating and worrying field of oncology and cardiology (López-Sendón et al. 2020).

Indeed, among cancer patients, the prevalence of cardio-vascular (CV) events is many times higher than that in the general population. Widely used anti-tumor drugs, such as anthracyclines (e.g., doxorubicin) and alkylating drugs (e.g., cyclophosphamide), have enormous multi-organ toxigenic potential, particularly cardiotoxic. Yet, several other drugs may contribute to CV damage as analyzed in this book.

The most challenging problem facing cardiologists and oncologists is the decision about starting, continuing or chemotherapy in case of development of cardiovascular diseases.

This book analyzes some of the most burning questions in Cardio-Oncology.

In the various chapters of this book, in addition to in-depth discussions on pathophysiological mechanisms of cardiotoxicity, readers can also find therapeutic measures and strategies to limit cardiotoxic damage in clinical practice. We hope the readers can find this book useful for their daily work as researchers and/or clinicians involved in Cardio-Oncology.

References

1. López-Sendón J, Álvarez-Ortega C, Zamora Auñon P, et al. Classification, prevalence, and outcomes of anticancer therapy-induced cardiotoxicity: the CARDIOTOX registry. Eur Heart J. 2020;41(18):1720-1729. doi: 10.1093/eurheartj/ehaa006.

Index

About the Editors

Dr Valentina Mercurio got her MD in 2010, her Board in Internal Medicine in 2016 and her PhD in 2019 at Federico II University, Naples, Italy. She is currently Assistant Professor of Medicine in the Department of Translational Medical Sciences, Federico II University, Naples, Italy, since 2019, where she coordinates the Echocardiographic Laboratory of the Cardio-Oncology Unit and of the Heart Failure outpatient unit mostly dedicated to post ischemic heart failure, right ventricular dysfunction and pulmonary hypertension, and in particular to the follow-up of oncologic patient before, during and after undergoing antineoplastic therapies. Her previous studies and ongoing collaborations with Prof. Paul Hassoun and Prof. Monica Mukherjee at Johns Hopkins University, Baltimore, MD, USA are mostly focused on pulmonary hypertension and right ventricular dysfunction, and she has established herself as a Pulmonary Hypertension basic and clinical investigator in Naples. Since 2019, Dr Mercurio is Fellow of the Italian Society of Cardiology and nucleus member of the working group on drug-induced cardiotoxicity and cardio-protection, and a member of the ESC Council of Cardio-Oncology, member of the Heart Failure Association, and of the WG Myocardial Function and of the WG on Pulmonary Circulation & Right Ventricular Function of the ESC, and of the PVRI (Pulmonary Vascular Research Institute).

Prof Pasquale Pagliaro, M.D., Ph.D was born in Rossano, Italy, in 1961. He is a full professor of Physiology at University of Turin (Italy) Department of Clinical and Biological Sciences. He is also member of the National Institute for Cardiovascular Researches (Bologna, Italy).

Degrees awarded: MD, University of Turin (Italy), Thesis topic: Coronary Pathophysiology, 1988. PhD, University of Turin (Italy), Thesis topic: Endothelial Physiology, 1994. Research Fellowship in Medicine-Cardiovascular at the Johns Hopkins University Baltimore (USA); Research topic: Coronary and Endothelial Physiology, 1997–99. Research experience/ other activities: Prof Pagliaro is PI in a lab studying coronary physiology and pathophysiology, and cardioprotection. His recent research concerns endothelial factors and other endogenous substances in triggering protective signaling pathways. Prof Pagliaro's lab also focusses on redox-signaling and mitochondrial function.

Prof Pagliaro served as an Ordinary member of Italian Society of Physiology, The Physiological Society (London), Italian Society of Cardiology, European Society of Cardiology, Italian Society of Cardiovascular Research (SIRC). He served as Vice-coordinator of the nucleus and as member of the working group on drug-induced cardiotoxicity and cardio-protection of the Italian Society of Cardiology. Prof Pagliaro is Past-President of SIRC and the Coordinator of the PhD in Experimental Medicine and Therapy of the University of Turin.

Prof Claudia Penna, BSc.D., Ph.D was born in Asti, Italy, in 1967. She is an associate professor of Physiology at University of Turin (Italy) Department of Clinical and Biological Sciences. She is also member of the National Institute for Cardiovascular Research (Bologna, Italy).

Degrees awarded: BSc, University of Turin (Italy), Thesis topic: Effect of venom in the isolated heart, 1991. Specialist of Clinical Pathology, University of Turin (Italy), thesis topic: Modulation of cardiac current by Nitric Oxide, 1995, PhD, University of Turin (Italy), thesis topic: Hyperaemic response and Ischemic Preconditioning, 2000.

Research experience/other activities: cardioprotection and cardiotoxicity.

She is member of Italian Society of Physiology, the Italian Society of Cardiology and European Society of Cardiology. She served as Vice-coordinator of the nucleus and as member of the working group on drug-induced cardiotoxicity and cardio-protection of the Italian Society of Cardiology.

Prof. Carlo Gabriele Tocchetti got his MD in 1997, his Board in Cardiology in 2001 and his PhD in 2007 at Federico II University, Naples, Italy, and is currently Associate Professor of Medicine and Director of the Cardio-Oncology Unit in the Department of Translational Medical Sciences, Federico II University, Naples, Italy since 2014. He coordinates the Heart Failure outpatient unit mostly dedicated to post ischemic HF, RV dysfunction and Pulmonary Hypertension, and to the follow-up of oncologic patient before, during and after undergoing antineoplastic protocols, and has established himself as a HF basic and clinical investigator. Prof. Tocchetti is Fellow of the Heart Failure Association (HFA), 2020–2022 Chair of the Study Group on Cardio-Oncology of the HFA, of the HFA Translational Research Committee, and of the WG on Myocardial Function of the European Society of Cardiology (ESC), and 2020–2022 Board Member of the ESC Council on Basic Cardiovascular Science (CBCS) and Council of Cardio-Oncology (CCO). His previous studies and ongoing collaborations with Drs Kass and Paolocci at Johns Hopkins University, Baltimore, MD, USA, have helped dissecting the cardiac contractile effects of HNO and the development of a novel HNO donor for treating human heart failure currently used in clinical trials. Hence, the main goal of his lab is to explore pathophysiologic mechanisms and therapeutic targets in cardiac dysfunction, with a particular interest on post-ischemic HF, genetic and inflammatory cardiomyopathies, Pulmonary Hypertension and RV dysfunction, and HF due to antineoplastic therapies, including novel anticancer immunotherapies and biologic drugs employed in inflammatory diseases.

For Product Safety Concerns and Information please contact our EU representative GPSR@taylorandfrancis.com
Taylor & Francis Verlag GmbH, Kaufingerstraße 24, 80331 München, Germany

www.ingramcontent.com/pod-product-compliance
Ingram Content Group UK Ltd.
Pitfield, Milton Keynes, MK11 3LW, UK
UKHW022211190926
475702UK00003B/301
* 9 7 8 8 7 7 0 2 2 6 2 8 8 *